The Carbon Dioxide Revolution

Michele Aresta • Angela Dibenedetto

The Carbon Dioxide Revolution

Challenges and Perspectives
for a Global Society

Michele Aresta
LAB H124, Tecnopolis
Innovative Catalysis for Carbon Recycling
Valenzano, Bari, Italy

Angela Dibenedetto
CIRCC and Department of Chemistry
University of Bari Aldo Moro
Bari, Italy

ISBN 978-3-030-59063-5   ISBN 978-3-030-59061-1   (eBook)
https://doi.org/10.1007/978-3-030-59061-1

This Springer imprint is published by the registered company Springer Nature Switzerland AG
The registered company address is: Gewerbestrasse 11, 6330 Cham, Switzerland

"The art piece is a creation by Antonia "Tonia" Copertino, painter and Professor of Fine Arts living in Molfetta (Bari-IT)

Since the 1980s she has been an active scholar and interpreter of sign as a artistic language and she has been present in national and international exhibitions and galleries

Tonia here shows CO_2 converted into Cn energy rich molecules at a catalytic center"

To our Families

This Book was inspired by our thoughts about the future of our children and grand-children.

Which life will they live? Will they have a better life than we have or will they suffer limitations?

Will energy be plentiful and which forms of will they use? Will they use less fossil-C? Will they breath cleaner air?

We are both optimist and see blue sky and clean water for future generations.

Earth will recover, yes it will.

Nature will be able to cancel the signs left by man: it has shown its power in these days signed by COVID 19 that obliged a large number of people to stay home, use less cars in cities and outside, emit less pollutants.....

In a month, a different life appeared all around us: clean air, blue sky, limpid water in rivers and lakes...

The quality of our life depends on us.

We need to change our lifestyle, we need to be closer to Nature, using the best of innovation.

Preface

This book, intended for non-specialists, graduate, and undergraduate students, features the epic change in the attitude of policy-makers and industrialists toward carbon dioxide, CO_2, which is not anymore considered a *waste* but a *resource*. As a matter of fact, nature, in its carbon cycle, cycles CO_2 at a rate of over 220 Gt of C per year or over 800 Gt_{CO2}/y for making a myriad of energy-rich compounds, using water as source of hydrogen and sun for powering the various processes. But nature cannot buffer the 37 Gt_{CO2}/y emitted within anthropogenic activities.

The utilization of CO_2 as building block for chemicals and materials, and source of carbon for fuels is a strategy we have beaten the drum for since early 1980s. So far, there were not the conditions for its large-scale implementation. These days, the need to *shift from the linear to the circular economy* and the exploitability of perennial primary energy sources (sun, wind, hydro, and geothermal energies) have brought to the attention of policy-makers and industrialists the great opportunity offered by using carbon dioxide. The use of CO_2 avoids the extraction of fossil natural resources and the transfer of carbon from the ground to the atmosphere, which is considered responsible for *climate change*. Is CO_2 the *protagonist* of climate change we are observing these days? Opinions about the real role of CO_2 are presented in this book: maybe it has a deuteragonist or even lower role. An analysis is made of primary energy sources used today and of the perspective huge change underway, with fossil-C expected to go down from actual 80.2% of share as primary energy source to a possible 25% by 2040, the difference being covered by perennial energies and renewable energy such as biomass.

The role of fossil-C in the actual energy system is analyzed (Chap. 1), completed with the use of fossil-C in the chemical industry (Chap. 2). An analysis is made on the impact of such use on the natural C-cycle and the increase of the natural greenhouse effect (Chap. 3). The role of efficiency strategies in the production and use of energy (electric, thermal, and mechanical) in our world and of renewable-C (biomass) is discussed as a way to the reduction of the emission of CO_2 (Chap. 4). This key topic is followed by the analysis of the potential of alternative C-free energy sources (Chap. 5) and of the capture and storage technology for CO_2-sequestration underground or into long time span materials, such as inorganic carbonates (Chap. 6). The electronic and physical properties of CO_2 are presented in Chap. 7, that is, the prologue to the use of CO_2 as technical fluid (Chap. 8) or as building block for chemicals and materials and as source of carbon for fuels (Chap. 9), emphasizing why so far the large-scale exploitation of such option was not realistic. Solar chemistry for CO_2 conversion into energy-rich products is featured in Chap. 10, while integrated chemo-biosystems are described in Chap. 11. Chapter 12 opens the window on our future and on large-scale perspective CO_2 utilization, presenting figures that indicate how our future will change.

A carbon dioxide revolution implements the natural C-cycle into industry, because CO_2 is renewable carbon. Using CO_2 means to avoid the extraction of fossil-C and its transfer to the atmosphere. The implementation of the Carbon Circular Economy-CCE principles in human activities and C-management, coupled to the exploitation of the bio-economy concept, will improve the quality of our life and save natural resources for next generations, while offering the opportunity for increasing the standard of life all over the planet.

Valenzano, Italy   Michele Aresta
Bari, Italy   Angela Dibenedetto

Contents

Energy and Our Society

Abstract

Fossil-C, in the form of coal, natural gas, and oil, covers 81+% of the energy necessary to satisfy the needs of our society. The continued use of fossil-C causes the accumulation of carbon dioxide in the atmosphere. The reduction of the emission of CO_2 is becoming urgent.

1.1 Introduction

Humans, as all living organisms, emit bio-CO_2 during the expiration (*ca.* 0.9 kg/day person, or 328 kg/y person, or *ca.* 25 t/person in a life of 75 y). Therefore, over 7 billion people in the actual world population emit *ca.* 7 Mt_{CO2}/day or 2.55 Gt_{CO2}/y just for the life's very basic operations such as respiration.

Since man uses fire, it has also become an emitter of CO_2 produced through the combustion of fuels in a variety of activities, including cooking. During 2018, the total World Energy Consumption (WEC) was of 13 978 Mt_{oe} (Mt_{oe} = million-ton-oil-equivalent, i.e., expressing all the energy used as oil), [1] continuing the growth observed over the last 28 years (Table 1.1).

Most of the total energy (81+%) comes from fossil-C-based sources (coal, lignite, oil, gas), which are converted into other forms of energy such as thermal, electrical, mechanical, etc. with a quite low efficiency in the range 27–50% and an average of *ca.* 30–33%. Major users are the production of electric energy, industries and transport. The remaining part of the original chemical energy of fossil-C is lost as heat, often at high temperature, that ends with a direct heating of the atmosphere and our Planet.

© Springer Nature Switzerland AG 2021
M. Aresta and A. Dibenedetto, *The Carbon Dioxide Revolution*,
https://doi.org/10.1007/978-3-030-59061-1_1

Table 1.1 Total energy consumption as $Mt_{oilequivalent}$ and its segmentation per areas on earth

Year	Middle East	Africa	Asia	Pacific	Latin America	North America	CIS	Europe	Total
1990	223	381	2 113	104	462	2 121	(1 372)	1 785	8 561
2000	372	487	2 886	129	598	2 523	898	1 853	9 746
2010	647	689	4 825	150	784	2 481	1 008	1 927	12 511
2018	803	850	5 859	158	822	2 558	1 081	1 847	13 978
2018% increase with respect to 1990	260	123	177	52	78	21	(−21) 20	3.5	63

Note Only CIS has apparently shown a decrease of energy consumption (figures in parentheses) over the period 1990–2018 during which a variation of composition of the Confederation occurred. On the other hand, if calculated over the period 2000–2018, an increase of 20% in energy consumption for CIS is observed. The highest increase is observed for developing countries, with India, China, Middle East and Africa leading the world growth of energy consumption

1.2 The Fossil-C-Based Energy Frame

Carbon-based fuels have been used by mankind as source of heat and energy, in general, since the first anthropogenic fire (Fig. 1.1). Man-controlled fire has a life of *ca.* 1.5 My.

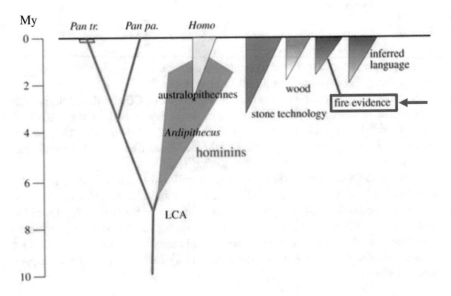

Fig. 1.1 The appearance of Hominins and Homo and the first anthropogenic fire. *Reproduced from Ref* [2] *(CC BY 4.0). Pan Tr. = Pan Troglodytes; Pan pa. = Pan Paniscus; LCA = Last Common Ancestors of Hominins and Pan*

Biomass	Coal	Gas	Oil
1.5 My	3490 bC	1000-500 bC	600 bC

Scheme 1.1 The use of C-based fuels over the time

Controlled fire has permitted the development of a quite large number of applications during millennia (heating, cooking, metal forging, electricity generation, transportation, etc.). Slowly, man moved from wood to coal, natural gas, and finally oil, always using C-based fuels (Scheme 1.1).

1.2.1 A Short History of Coal

Coal was discovered around 3490 bC by Chinese people who soon started to make a household use of it. Later on, also Greeks and Romans started to use coal for diverse applications. Curiously, its brightness has suggested the use in jewellery and pieces of coal were used as pendants in necklaces or cufflinks by Aztecs, the earliest use of coal in Americas.

Nativity Handcrafted from Coal (Kentucky, USA)

Still today, coal is used for making personal ornaments. Romans were the first to discover and extract coal in Britain, where already during the first two centuries after Christ coal was largely employed for many uses, including forging metals. The blooming of the Industrial age was possible thanks to coal. At the beginning of 1800, the countries today known as UK and USA were the largest producers and marketers of coal with 260 and 350 Mt/year, respectively. Since then, coal has been extracted all over the world where it is distributed with different abundance among countries. Australia, North America, Asia, Europe, and Africa are rich of coal that for decades has been the main source of energy.

W1.1 Average dry atmosphere composition and its pollution

Dry atmosphere, the gaseous mass that surrounds our Planet Earth, is composed of 78.08% N_2, 20.94% O_2, 0.93% Ar, 0.04% CO_2, 0.00005% He, and other trace gases. Figures given above represent an average composition, which slightly varies with the height over the Earth surface: heavier gases (CO_2, Ar, O_2) are more abundant close to the surface, lighter (He) in the higher layers. Water vapor can be present in various amounts (humidity), depending on the climatic conditions. Emissions by fossil-C combustion cause accumulation in the atmosphere of pollutants that may cause damages to goods and humans. For example, the emission of S or N oxides (SOx, NOy) causes the formation of acid species which can fall down with rain (*acid rain*) and cause serious damages to humans, vegetation and even buildings.

Some apparently inert species can be converted by sunlight in the atmosphere and generate dangerous pollutants. Other gaseous species, like chlorofluorocarbons (CFCs) (vide infra), can pass the troposphere unaltered (Chap. 2) and reach the stratosphere where they are converted by UV solar radiations into species which destroy *ozone* and cause the known *ozone hole*. Particulate of various dimensions can be suspended in the atmosphere and reach a variety of targets. Inspiration of very fine particulate (PM2.5 and PM5, dimensions of 2.5 and 5 micrometers) is very dangerous as it can reach deep parts of the respiratory apparatus and cause cancer.

The actual worldwide production of coal is summarized in Table 1.2: it amounts to *ca.* 7 732 Mt/y, including lignite. China is the largest producer and user of coal, facing quite heavy environmental problems. As a matter of fact, coal is a strong emitter of pollutants when burned, because it contains sulfur that generates SOx and metals (even some toxic metals) that accumulate in ashes, in general, as oxides. During the combustion, a fine particulate is emitted that causes deep atmosphere alteration (**W1.1**). During the great coal age (1800s–1900s), when coal was largely used for heating civil buildings, smog was wrapping large cities (famous is the London gray-greenish atmosphere, after which a color was named, the *London gray*). After 1970s, the growth of the environmental consciousness pushed toward the cleaning of coal used in heating of civil buildings in cities and in industrial processes. Deep pre-treatment of coal (*desulfurization*) and abatement of particulate were implemented over a large scale, or even coal was substituted with cleaner fuels

Table 1.2   Worldwide production of coal

Country	China	India	USA	Russia	Germany	Japan	South Africa	South Korea	Poland	Turkey	Australia	Indonesia
Mt/y	3 770	982	624	234	217	189	186	150	129	125	113	109

(*natural gas*) so that the situation is today much improved, even if largest cities, located in not well-aerated areas, have still to face critical situations. Technology innovation has made that the direct burning of coal is even avoided and instead it is converted into other forms of cleaner fuels, mostly through its conversion ("*re-forming process*") into syngas—a mixture of CO and H_2—which is converted through the catalytic Fischer–Tropsch (FT) process into gaseous, liquid, and solid products (see Chap. 2). South Africa and Malaysia make most of their gasoline through FT processes. The application of technologies such as Water–Gas Shift (WGS) to reforming allows to convert coal and water into H_2 and CO_2 (Eq. 1.1a–1.1b) that are separated and H_2 can then be used in a clean combustion with oxygen affording water as combustion product. (Eq. 1.2). If such reaction takes place in a cell (*fuel cell*), electricity is produced.

$$C + H_2O \rightarrow CO + H_2 \rightarrow (syngas) \qquad (1.1a)$$

$$CO + \mathbf{H_2O} \rightarrow CO_2 + H_2(\text{WGS}) \qquad (1.1b)$$

$$H_2 + 1/2O_2 \rightarrow H_2O + \boldsymbol{Energy} \qquad (1.2)$$

This approach is at the basis of the innovative technology known as Integrated Gasification Combined Cycle (IGCC) that allows decarbonization of fuels with CO_2 capture prior to combustion for making electric energy (see Chap. 2).

1.2.2 Natural Gas: Its Discovery, Early and Actual Uses

Natural Gas (NG) is known since 1000 bC and was discovered in China. Spontaneous emissions were used as flames in front of or inside temples by antic Greeks. Such emissions were also known as "*eternal fires*," mostly used as house ornament. Chinese people first tried to transport NG using bamboo pipes. In 1800, NG was transported using cast iron pipelines and largely used for lightening houses and streets (UK, since 1785, and USA were pioneer in such application). The discovery and exploitation of oil pushed NG to a second level of importance because of the easier transportation of oil and its higher energy density. Today, NG is coming back to massive use because it emits less CO_2 than oil and coal for the same energy generated. In fact, for producing one kWh of electric energy, *ca.* 1 kg of CO_2 is produced burning coal, while only 0.5 kg_{CO2} are emitted using NG, with oil sitting in the middle. NG is considered a clean fuel for use in houses (heating, cooking), transport (cars), industry (process powering), and electricity production. Table 1.3 gives the worldwide production of NG.

Table 1.3 Worldwide yearly production of natural gas as billion cubic meters (bcm)

Country	Europe	Asia	North America	Latin America	Pacific	Middle East	CIS	Africa
Bcm	538	787	976	243	50	539	671	151

NG, which is made mainly of methane, is also used in the chemical industry as starting material for producing chemicals (see Chap. 2) through its conversion into syngas (*methane reforming*). The recent economic implementation of the technology of "*shale fracking*" (hydraulic fracturing of shale clay rocks that contain gas formed upon decomposition of organic materials) has opened a new market that is already exploited at an interesting level in USA.

1.2.3 Oil and Its Superior Properties

The discovery of oil dates back to 650 bC in China where first attempts to transport by pipelines (made of bamboo) were also made. It had a spot utilization until 1859, when it was discovered and drilled in Pennsylvania, USA. With the discovery of the Texas-USA wells in 1901, the large-scale commercialization of oil started. Today, oil is drilled all over the world not only on Earth surface, but also in seas and oceans, reaching deepness (7 000$^+$ m) not imagined 40 years ago. Oil is the main source of fuels and chemicals today. On oil refining and conversion is based the largest chemical industrial sector: petrochemistry. The total production of oil today is reported in Table 1.4.

Oil is the most concentrated form of solar energy. Table 1.5 reports the energy density of several vectors. Long-chain liquid hydrocarbons distilled from oil are by far the most efficient energy carrier, better than coal, NG, methanol and much better than H_2 and batteries. This property makes oil the most suited source of fuels for cars and several other applications. Oil is easy to ship everywhere and can be easily stored in tanks without problems. If it is carefully handled and used, it does not create any problem to the environment and health. For this reason, oil finds ubiquitous use and is difficult to substitute in transport and other applications.

Table 1.4 Worldwide production of oil in Mt/y 2018

Country	Europe	Asia	North America	Latin America	Pacific	Middle East	CIS	Africa
Mt/y	168	349	935	432	Na	1 496	691	398

Table 1.5 Energy density of various vectors

Vector	Diesel	Gasoline	Carbon-coke	Brown coal	Methanol	Bio-oil (algae)	H_2(l)	CH_4 (g) c	H_2(g) 20 MPa
Density GJ/m^3	36	34	30	18	17	13	9	8	2

1.3 The Fossil-C Availability

Fossil-C is, in principle, renewable, but, unfortunately, it cannot be regenerated at the rate we use it: its availability is got to inexorably decrease with time. This is due to the difference between the rate of combustion of biomass and that of biomass generation: the former is some 1 000–10 000 times faster than the latter. In fact, the rate of production of CO_2 by combustion of carbon-based materials ($6 - 15$ g_{CO2} m^{-2} s^{-1} for several kinds of wood [3]) is much faster than the fixation of CO_2 by photosynthesis (the best values are observed in microalgae [4], more efficient fixing agents than terrestrial plants, in the range $0.1–1.7$ g_{CO2} $L^{-1}day^{-1}$ or $0.00057 - 0.0098$ g_{CO2} $m^{-2}s^{-1}$). Additionally, the increasing population (our planet population is 7 Bpersons today and is expected to grow to 10 Bpersons by 2030), the ethic question of assuring a decent quality of life to 1.1 Bpeople (UNICEF 2017) who live in poverty (with 750 Mpeople living in extreme poverty), and the general wish of improving the worldwide average standard of life demand more energy.

In several areas, deforestation is pervasive (land is used as arable soil or for setting new production sites or inhabitations): *the agents able to fix CO_2 (trees) are decreasing in number while the emitters (humans) are increasing with time, reinforcing the CO_2 accumulation in the atmosphere.*

However, our society is squeezed between two walls: from one side the increasing need of energy for satisfying wishes of a high standard of life of an ever-increasing number of persons and from the other the impossibility of satisfying such necessity in the usual way (use of fossil carbon), because of the scarce availability for the future generations and the putative negative effects of the emission of CO_2 on climate.

For how long can we rely on fossil carbon, then? Table 1.6 gives an estimate of the reserves of fossil carbon per geographic area. The distribution is important because political factors may affect the general usability of resources.

The total amount of extracted fossil-C in 2018 was categorized below:

- *Oil: 4 472 Mt_{oil},*
- *Coal and lignite: 7 732 $Mt_{coal\ and\ lignite}$, and*
- *Natural gas: 3 955 Bm^3 of natural gas.*

Not all such carbon was used. For example, oil was used at a rate of 4 300 Mt in 2018.

Table 1.6 Estimated world fossil-C reserves (Gt_{oe})

Region		Europe	Russia	N. America	S. America	China	India	Middle East	Africa	Australia	Total
Fossil-C reserves, Gt_{oe}	Coal	40	152	170	13	76	62	0	34	60	607
	Oil	2	19	8	15	2	1	101	17	2	167
	Gas	6	52	8	6	2	1	68	14	10	167
	Total	**48**	**223**	**186**	**34**	**80**	**64**	**169**	**65**	**72**	**941**

Table 1.7 CO_2 emission in the combustion of various C-based fuels (natural or man-made) for producing 1kWh (1 GJ, Column 3) of energy

Fuel	Emis-sion in kg_{CO2}/kWh	Emis-sion in kg_{CO2}/GJ
1. Wood *)	0.39	109.6
2. Peat	**0.38**	**106.0**
3. Lignite	0.36	101.2
4. Lusatia	**0.41**	**113.0**
5. Central Germany	0.37	104.0
6. Rhineland	**0.41**	**114.0**
7. Hard coal	0.34	94.6
8. Fuel oil	**0.28**	**77.4**
9. Diesel	0.27	74.1
10. Crude oil	**0.26**	**73.3**
11. Kerosene	0.26	71.5
12. Gasoline	**0.25**	**69.3**
13. Refinery gas	0.24	66.7
14. Liquid petroleum gas	**0.23**	**63.1**
15. Natural gas	0.20	56.1

*) Commercial quality pellets are considered as a widespread thermal energy source. See Chap. 4

When burned, fossil fuels and biomass all produce CO_2 and the total amount of emitted CO_2 by fossil fuels use was 32 000 Mt_{CO2} in 2018. Fossil-C is used at a large extent for producing electric energy. Different C-fuels have different specific capacities of producing electricity and cause different specific emissions. Table 1.7 shows how differently various C-sources behave with respect to the production of a given amount of electric energy (1 kWh, Column 2, or 1 GJ, Column 3) and the relevant CO_2 emission: if compared to natural gas (last entry in Table 1.7), oil (Entry 8) emits 40% more CO_2, hard coal (Entry 8) 70%, and lignite (Entry 3) 80%. Peat and wood are still worst. The use of biomass as source of energy is becoming quite appealing with respect to fossil-C because biomass is formed from atmospheric CO_2 and, in principle, would produce energy at quasi-zero CO_2 emission. Indeed, the zero-emission option is not feasible because even using spontaneous natural biomass, energy is used in the collection-transport-processing of biomass that must be taken into account.

Different biomasses have different energy contents, much lower than fossil-C, and emit larger quantity of CO_2: humanity today and in future cannot leave on biomass as the only source of energy (See Chap. 4).

Table 1.8 reports the capacity of biomass power plants in selected countries and worldwide in 2018 (in gigawatts) [5]. As capacity is intended the total installed power, even if not really used.

Currently, biomass covers approximately 10% of the global energy supply, of which two-thirds are used in developing countries for cooking and heating. In 2009, about 13% of biomass use was consumed for heat and power generation, while the industrial sector consumed 15% and transportation 4% [6].

Table 1.8 Capacity of biomass power plants in selected countries, GW

Country	US	China	India	Germany	UK	Japan	Total
Capacity (gigawatts)	16.2	17.8	10.2	8.4	7.7	4	130

Another point of consideration is that direct burning of biomass is not a clean process: it has a strong environmental impact in terms of emitted particulate, unburned materials, N-containing organic pollutants derived from wood-endogenous-N, dioxins, etc. For the large-scale combustion of biomass, especially in large cities, clean technologies are needed that prevent atmospheric pollution.

The benefit is in the fact that biomass is produced from atmospheric CO_2: the International Panel on Climate Change (IPCC) estimates that the modern biomass usage has a large carbon mitigation potential. The mitigation potential for electricity generation from biomass will reach 1 220 Mt_{CO2eq} for the year 2030; a substantial fraction of it at costs lower than 19.5 $US\$_{2005}/t_{CO2}$.

The massive use of pellets for heat generation in urban areas is causing the formation of heavy smog. In order to reduce the environmental burden, it is necessary to process biomass into clean fuels before using it: this will rise issues about its economic and energetic cost. Even if the use of renewable sources (biomass) and perennial energies (solar, wind, geo, hydro-SWGH power) will be expanded, it is unlike that perennial and renewable sources will cover the need of energy of humanity. *The use of fossil-C use will be progressively reduced up to minimize it, but will not reach zero.* Most likely, fossil-C will still be necessary for feeding high-density and high-intensity uses, such as heavy electric terrestrial transport, naval transport and aeronautics, and as raw material for the chemical industry (that cannot be decarbonized), while perennial and renewable sources will likely be used for low-density and low-intensity uses (domestic uses and some kind of light transport). However, the combustion of fossil-C is causing the emission into the atmosphere of large amounts of CO_2 which accumulates and rises serious concerns about its putative impact on climate change. Therefore, the reduction of CO_2 emission is a must for our generation.

This urgency is aggravated by the fact that natural resources are not infinite: even if the availability reported in Table 1.6 can somehow be expanded by discovering new oil or gas fields or coal mines and by developing technologies that will allow the exploitation of fields today not reached, it is a matter of ethics to save resources for next generations, limiting consumption today: they will not last forever. At the actual rate of consumption, we have enough fossil-C for only 70 years. Our dilemma is that our society demands more energy, while we should use less C-based primary sources. How to match, thus, the request of more energy and the need to reduce the use of fossil-C that provides 81+% of the total energy used today? The most intuitive answer is: increasing the efficiency in the production and use of energy and exploiting alternative non-C-based energy sources. This is all? Any other innovative solution? Yes. A revolutionary approach to problem-solving based on "*Carbon Recycling-CR*" is growing, based on technology innovation, system integration, and coupling chemistry–catalysis–biotechnology. *Carbon recycling merges the intensity typical of man-made (or industrial) processes with the Nature-inspired "cooperative-systemic-cyclic" concepts. The aim is a "Man-made C-cycle" that may enhance the rate of CO_2 conversion with respect to natural processes, and its integration with the natural C-cycle for producing goods and fuels: recovery and reuse of carbon is a new paradigm in the CO_2 problem.* It is our firm belief that such a complex problem, namely, the CO_2 emission reduction, cannot be solved by a single option exploitation, and requires instead an *integrated solution* and *C-recycling is a strong part of it.*

1.4 Recovery and Reuse of Goods

Recycling of goods is a practice applied to several materials since very long time: it is time now that we apply the same concept to carbon. Metals (aluminum, copper, iron, gold, silver, and many others) are recovered from industrial slags and/or used products and reused in order to save natural resources and, in some cases, prevent pollution and save energy. Also, municipal or industrial wastewater is treated, sanitized, and reused at large extent in some geographical areas. Glass is recovered at the end of life of bottles and other goods and reused. Paper is recovered and reused. It is becoming more and more imperative these days to recover and reuse plastics.

Now, it is time that we learn to efficiently recover and reuse carbon. Here is our future.

Table 1.9 gives an idea of the nature and percentage of goods that are recovered and recycled. It is worth to mention that recycling of a given good can be performed either in the same production cycle (primary recycling) or in a different process that produces goods of lower quality and use. For example, plastic used for food packaging once recovered and recycled most likely will not be used for producing the same quality plastics because of potential pollutants that are incorporated in it.

Unless food-level purity is matched, recovered plastics will be used in a lower level application: they will be suited, thus, for producing items not used in the food sector, such as materials for industrial applications or pipes for irrigation. Medical plastics are not reused as they can be carriers of infectious microorganisms and cells.

Reuse requires a lot of care, for avoiding that health of humans may be affected: this is true for all goods, including carbon. Recovered (from power plants or industry) CO_2 can find use in several non-chemical applications (see Chap. 8), including additive to beverages, preservative of food, and modified packaging: only food-grade CO_2 will be used in the three latter applications. Table 1.9 shows that carbon (in the form of coal, oil, gas, and biomass) is by far the most used good, but also the one that is less recycled (as %), as for now. The reasons why will be discussed in following chapters and how such situation is changing will be described.

Table 1.9 Some hints in recycling of goods in anthropic activities

Good	Worldwide used volume/mass	% Recycling	Notes
Freshwater (global: agricultural, industrial, and drinking) [7]	3 996 757 700 000 m^3/y Per capita: largest user Turkmenistan 16 281 L/d; lowest: DR Congo 34 L/d Average use: 70% agriculture, 20% industry, 10% drinking. In industrialized countries, the industrial use of water can reach 80% (Belgium)	< 1 A target of 30% worldwide average recycling is set for 2050	Recycling best cases: Israel 85%; Kuwait 35%; Singapore 30%; Queensland (AUS): > 35%; California (USA): 50% in Orange Country
Paper	**423 Mt/y (2017)**	*Ca.* **50%** **Recycling paper saves 65% of the energy needed to make new paper and also reduces water pollution by 35% and air pollution by 74%**	**Recycling one ton of paper saves up to 31 trees, 4 000 kWh of energy, 1.7 barrels (270 L) of oil, 10.2 million Btu's of energy, 26 000 L of water, and 3.5 m^3 of landfill space**
Iron	1 000 Mt/y	40	Iron is the good recycled since longer time
Aluminum [8]	*ca.* **94 Mt/y (production 2017)** *ca.* **81 Mt/y (used 2017)**	31 In special cases, Al is made of 75% recycled material	**Aluminum is one of the most commonly recycled goods**
Copper	19.1 Mt/y (2017)	> 45	Copper can be recycled without loss of properties
Plastics	**360 Mt/y**	9	**Recycling depends on the composition and structure of the polymeric material**
Glass	205 Mt/y	33.9 Six tons of recycled glass avoid 1 t of CO_2 emission	The use of recycled glass depends on the properties of the recovered materials. Food class glass can be reused for the same use only if sorted in the proper way
Fossil-C Produced CO_2	**13.4 Gt/y (as C)** **32.4 Gt/y**	**0.68**	**Used mainly (90%) in the chemical industry. The rest is used for technological uses, including EOR**

References

1. https://yearbook.enerdata.net/total-energy/world-consumption-statistics.html
2. Gowlett JAJ (2015) The discovery of fire by humans: a long and convoluted process. Phil Trans R Soc, B37120150164
3. Tran HC, White RH (1992) Burning rate of solid wood measured in a heat release rate calorimeter. Fire Mater 16:197–206
4. Adamczyk M, Lasek J, Skawinska A (2016) CO_2 biofixation and growth kinetics of Chlorella vulgaris and Nannochloropsis gaditana. Appl Biochem Biotechnol 179:1248–1261
5. https://www.statista.com/statistics/264637/world-biomass-energy-capacity
6. Schill SR (2013). http://biomassmagazine.com/articles/9444/iea-task40-biomass-provides-10-percent-of-global-energy-use
7. https://www.worldometers.info/water/
8. http://www.world-aluminium.org/statistics/massflow/

Fossil-C Application in the Energy and Chemical Industry

Abstract

In this chapter, the use of fossil-C as source of energy (compared to other sources) and as raw material for the chemical industry is discussed. Some fundamental processes are illustrated for the production of liquid fuels from either coal or natural gas. Emissions of CO_2 are quoted for several sectors. The syngas and Fischer–Tropsch processes are discussed more in detail in the Deep Insight (DI).

2.1 Fossil-C as Primary Source of Energy

The global consumption of energy was 13.97 Mt_{oe} at the end of 2017 [1] with *ca.* 40% increase with respect to 2010 (10 024 Mt_{oe}). Table 2.1 gives the sharing among different sources.

Table 2.1 is quite illustrative of the role played by fossil-C in our energy system and of the existing trend toward the exploitation of alternative primary sources. As for the last 30 years, fossil-C has stably provided over $81^+\%$ of the total energy consumption [81.19% (1990)–81.21% (2017)] with a net trend toward an increase of use of fossil-C, and of coal, in particular, +83%, despite international programs claiming CO_2 emission reduction. Renewable or biomass (grown and waste) is apparently growing (902 367 in 1990 and 1 329 064 in 2017) but its share has decreased from 10.29% in 1990 to 9.51% in 2017. Hydroelectric has grown from 2.1 to 2.51%, nuclear has apparently grown but its share is gone down from 5.99 to 4.92%. Only perennial primary energy has sensibly increased its share from 0.042 to 1.8%. Such trend demonstrates that the use of fossil energy permeates our life and its substitution is not easy. The expansion of the use of perennial energy can reduce the use of fossil energy, and this requires the realization of particular conditions (vide infra). Renewable energy can play a role, but limited by the fact that

M. Aresta and A. Dibenedetto, *The Carbon Dioxide Revolution*,
https://doi.org/10.1007/978-3-030-59061-1_2

Table 2.1 Sharing of energy generation among primary sources (*Perennial* indicates solar-wind-hydro-geothermal sources, *Renewable* indicates biomass-based sources, including waste biomass)

Year	Oil	Renewable	Hydro	Nuclear	Perennial	Gas	Coal	Total
1990	3 232 737	902 367	184 102	525 520	36 560	1 663 608	2 220 466	8 765 360
2000	3 662 674	1 011 986	224 693	675 467	60 054	2 072 291	2 316 665	10 023 830
2017	4 449 499	1 329 064	351 029	687 481	256 830	3 106 799	3 789 934	13 970 636

Units are in tons oil equivalent: t_{oe}

biomass growth rate is much lower (from 1 000 to 10 000 times [2]) than its combustion. Nuclear will not expand its potential until safety issues will not find a solution that may assure people against radiation damage in case of accident. However, a limited number of options are available to supplant fossil-C as primary source of energy and not all of them are suited for large-scale exploitation at high intensity. In our perception, the use of fossil-C in energy production can be progressively reduced in time, even quite drastically, but not eliminated. Our estimate is that it can decrease from actual 81.2% to 25–30% of the total by 2040–2050. Supplanting fossil-C in sectors, which require low-intensity and low-density energy will be much easier and rapid than in sectors requiring high-density and high-intensity energy. The rapidity of supplanting fossil fuels will be determined by how fast will be the exploitation of technology innovation and of perennial sources (S-W-G-H). There are several barriers to such epochal change as categorized below:

1. Most perennial sources are not ubiquitous.
2. Solar energy is ubiquitous but solar radiation is not the same at all parts of our planet.
3. Perennial energies such as S-W-H are discontinuous.
4. Fossil-C in all its forms can be easily shipped through the globe and can reach any place for local specific utilization, wherever the users are located.
5. Alternative primary energies (S-W-G-H) are local in exploitation and cannot be shipped. Only the product of their use (electricity) can be shipped, with limitation to distance for avoiding large losses. Electricity transfer undergoes 8–18% loss from the power plant to the consumer in actual networks.

However, the substitution of fossil-C requires robust innovative technologies and large investment.

2.2 Fossil-C as Source of Carbon in Synthetic Chemistry

Fossil-C is used since the beginning of the industrial age in synthetic chemistry where it finds a variety of applications.

2.2.1 Use of Coal

Coal is used since the middle 1800s as source of carbon in synthetic chemistry, with an apparent decline in the middle 1900s due to the competitive price and easier use of oil and natural gas. As a matter of fact, for 100 years from 1850 to 1940, most synthetic chemicals were produced from coal through the *"coal chemistry."* From 1940s onward, the *"coal chemistry"* was progressively substituted by *"petro-chemistry,"* based on use of oil and even methane. The reduction of the use of coal was justified by several factors, such as the (i) competitive cost of oil and gas; (ii) easier extraction techniques; and (iii) cleaner processes for hydrocarbons work-up with respect to coal. As a matter of fact, environmental concerns, having issues both in by-products produced and the effects of coal mining/extraction, were a potent driver for the shift from coal to hydrocarbons, liquid and gaseous. Recently, attempts to develop clean technologies for coal conversion have been made. The use of coal had recently a revival, but still it is under severe control due to the emissions. The rise of price of oil and gas will eventually bring back coal to the chemical industry? This question will remain open until clean technologies are developed. Countries rich of coal, and poor of oil and gas, have to weigh the cost of investments for safe extraction–conversion of their own coal with respect to external debt if they will use oil and gas purchased on the international market. The main use of coal in the chemical industry is for producing syngas (CO+H$_2$) through the Water–Gas Reaction (WGR) and Water–Gas Shift Reaction (WGSR) shown in Eq. 2.1a and 2.1b, respectively. The overall reaction is shown in Eq. 2.1c.

$$C + H_2O \rightarrow CO + H_2 \,(\text{syngas production}) \qquad (2.1a)$$

$$CO + H_2O \rightarrow CO_2 + H_2 \,(\text{water-gas shift reaction}) \qquad (2.1b)$$

$$C + 2H_2O \rightarrow CO_2 + 2H_2 \,(\text{overall reaction}) \qquad (2.1c)$$

By combining reactions 2.1a and 2.1b in the right ratio it is possible to produce syngas **[DI2.1]** with a variable H$_2$:CO ratio, from 1 to 2 and even higher values. Such syngas is used for the synthesis of methanol (CH$_3$OH) or, through the Fischer–Tropsch (FT) process, for the synthesis of gasoline, diesel, and other chemicals. Scheme 2.1 shows the many products that can be produced from methanol [3].

It also finds utilization in several other fields, such as fuel cells. (Appendix C) Coal clearly meets the interest of several industrialized countries, which are rich of coal mines (such as USA, Australia, Poland, China, India, Malaysia, and South Africa) and already have on stream an advanced FT technology. FT plants for Syngas to Liquid (SGtL) may have a size of the order of 330 000 t/y (South Africa) up to 1 Mt/y. Catalyst improvement for better specification of SGtL products (gasoline fraction, mainly, with gaseous and wax fractions minimization) is a continuous development because of the relevant economic interest: improving by

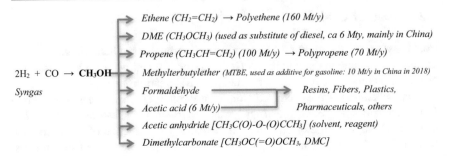

$2H_2 + CO \rightarrow$ **CH₃OH**

Syngas

Ethene (CH₂=CH₂) → Polyethene (160 Mt/y)

DME (CH₃OCH₃) (used as substitute of diesel, ca 6 Mty, mainly in China)

Propene (CH₃CH=CH₂) (100 Mt/y) → Polypropene (70 Mt/y)

Methylterbutylether (MTBE, used as additive for gasoline: 10 Mt/y in China in 2018)

Formaldehyde ────────── *Resins, Fibers, Plastics,*

Acetic acid (6 Mt/y) ──── *Pharmaceuticals, others*

Acetic anhydride [CH₃C(O)-O-(O)CCH₃] (solvent, reagent)

Dimethylcarbonate [CH₃OC(=O)OCH₃, DMC]

Scheme 2.1 Chemicals produced from methanol, their derivatives, and market volumes. Derivatives can also be obtained by other routes

1% the gasoline fraction yield means earning million dollars per year considered the global capacity of 38 000 m³/d or *ca.* 12 Mt/y.

Noteworthy, coal tar, a product of coke ovens, is a source of chemicals such as anthracene, naphthalene, phthalic anhydride, or carbon black, among others.

The *on-site* gasification of coal has been considered to be a major potential "on-purpose" source of commodity petrochemicals. Attempts to in situ (coal mine) gasification of coal into CO have been made in several countries, including Germany, but such approach has never reached a level of large exploitation (Scheme 2.2).

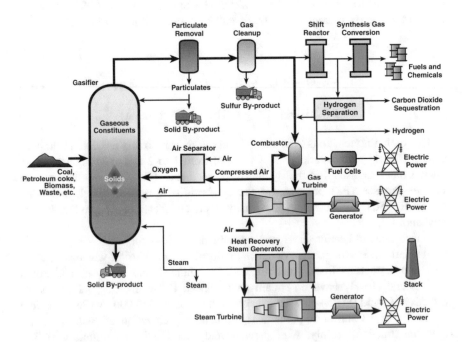

Scheme 2.2 Representation of the coal chemistry. Reproduced from Ref. [4] (CC BY 4.0)

Another "on-purpose" route for coal to chemicals is via the production of acetylene (HC≡CH), which can be used to produce a variety of chemicals, including vinyl chloride monomer (CH$_2$=CHCl) from which the polymer polyvinyl chloride (PVC) is derived. The economics for this production becomes attractive as the crude/coal price spread increases, and this route may prove important in areas with large coal reserves.

2.2.2 Use of Oil

Oil is used not only for producing gasoline and diesel for the transport sector, but is also the source of numerous valued chemicals, which are either present as such in the crude and can be separated or are made upon oil processing. Among them, we can cite gleaming paints, tough and moldable plastics, pesticides, detergents, cosmetics, pharmaceuticals, and so on. Industrial processes convert sluggish liquids into beauty products. By breaking the hydrocarbons chains into simpler compounds (CO) and then assembling those single-C building blocks, chemists have learned to construct molecules of exquisite complexity. Thus, destructive–reconstructive approach has been for long time the preferred approach to chemicals, and this approach suffers high energy expenditure due to the large entropy jump in going from structured C-molecules to de-structured single C-atom structures (Fig. 2.1). The overall transformation requires energy in the cleavage of C–C bonds, which is often lost in the global process.

Fig. 2.1 Entropy change in coal or hydrocarbons conversion into syngas (CO+H$_2$) and use of it in the synthesis of Cn chemicals

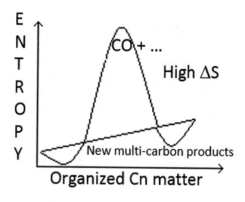

2.2.3   Use of Natural Gas

Natural gas (methane, CH_4) has a multitude of industrial uses, including providing the base ingredients for such varied products as plastics, fertilizers, anti-freeze, and fabrics. In fact, industry is the largest consumer of natural gas, accounting for 43% of natural gas use across all sectors.

$$CH_4 + H_2O \rightarrow CO + 3H_2 \qquad (2.2)$$

$$CH_4 + CO_2 \rightarrow 2CO + 2H_2 \qquad (2.3)$$

Natural gas as a feedstock is commonly found as a building block for methanol, which in turn has many industrial applications (Scheme 2.1). Natural gas is converted into syngas, through the catalytic Methane Wet Reforming (MWR) also known as Methane Steam Reforming (MSR) (Eq. 2.2). An alternative route is the so-called Methane Dry Reforming (MDR), which is based on the reaction of methane with CO_2, instead of water (Eq. 2.3). Natural gas is preferred to coal for the production of syngas because of less CO_2 emissions and lower environmental impact in general. In addition to these uses, there are a number of innovative and industry-specific uses of natural gas such as desiccant.

However, fossil-C is a basic component of the energy and chemical industry: supplanting it will not be simple, even if not impossible at least in some specific applications. The strategies will be discussed in following chapters, clarifying the role of CO_2 and renewable carbon in general.

2.3   Carbon Dioxide Emissions

All uses of C-based compounds will emit CO_2 at the end of the life cycle, as already reported. This is true for synthetic chemicals, vegetals, animals, and even in human life. Humans emit *ca.* 2.55 Gt_{CO2}/y or *ca.* 8% of the total amount emitted (Chap. 1). The emitted CO_2 reaches the atmosphere and enters the carbon cycle (see Chap. 3).

With the increase of population, use of energy and quality of life, the emission of CO_2 will continue to grow if there will not be a change in the actual trend. And such change must be based on fossil-C substitution or on carbon recycling, supporting Nature to make a wise use of CO_2.

As Fig. 2.2 shows, the world population is continuously growing even if there is a tendency to a reduced rate (Fig. 2.3). The consequent effect will be the increase of use of energy, even due to the general improvement of quality of life of humans. In a trend as usual, this will cause a serious increase of emission of CO_2 that will reach over 45 Gt/y in 2040.

It is believed that the reduction of the emission of CO_2 into the atmosphere has a key importance for keeping under control the climate change. Figure 2.4 shows the trend of anthropogenic CO_2 emission since the start of the industrial age. Until 1950

Fig. 2.2 Trend of world
population growth [5]

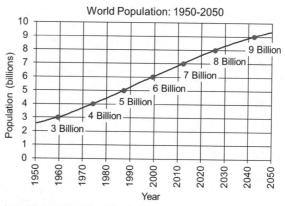

Source: U.S. Census Bureau, International Data Base, August 2016 Update.

Fig. 2.3 Trend of world
population growth rate [5]

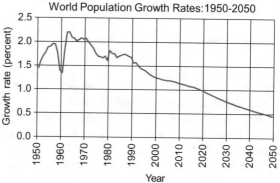

Source: U.S. Census Bureau, International Data Base, August 2016 Update.

the major emitters were USA and Western Europe, followed by Japan. The major expansion of the emission for such areas was during the 1950s–1980s when also Eastern Europe started to grow. Since then, an almost steady state was reached for the abovementioned areas, while starting the expansion of other areas such as China, India, Middle East, Africa, and other Pacific Countries than Japan. Recently, China has passed USA and is the major emitter of CO_2 worldwide.

Figures in Table 2.2 show that the production of electricity is the major emitter of CO_2, followed by transport and manufacturing industries. Some comments are appropriate in order to better understand the figures in the table. At the world level, the increase of emission for electricity production (line 1, column 2–3) tells about the spreading of such utility in areas not served so far. In fact, major increases are in the East Asia Pacific zone, Latin America, Middle East and North Africa, South Asia, and Sub-Saharan zone. Considering the partition by income, the increase is of

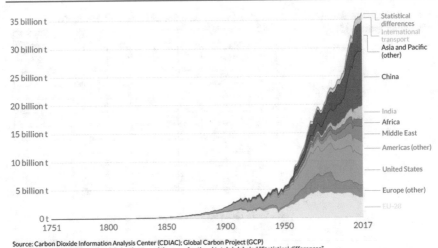

Fig. 2.4 Global emission of CO_2 by world region. Reproduced from Ref. [6] (CC BY 4.0)

34% for the low-income population, 12% for lower middle income, 16% for the upper middle income, and 8.8% for the high income. It is foreseeable that the major increase will be in the first two sectors even in future as the electricity will reach areas of our planet not yet served or poorly served up to now. A substitution of source, shifting from fossil-C to SWHG will be highly beneficial for reducing the CO_2 emission: in general such areas are well insulated and can make profitable use of solar energy.

Moving to the manufacturing industries and construction sector, one can observe a steady situation all over the world, with some reductions in some specific areas. This effect is the result of both technology innovation that saves energy and some shift to renewable energy sources.

The residential sector shows a stable emission in general. Other activities account for a minor, stable share of the emission.

Therefore, the sectors that need an immediate intervention are the production of electricity and, possibly, transport, which together account for 70% of the emission.

Shifting to PV electricity and biofuels may give a strong push to the reduction of the emission of CO_2. But why the CO_2 emission should be reduced and why is so urgent to reduce CO_2?

These issues will be discussed in Chap. 3.

Table 2.2 CO$_2$ emissions per world area and sector of activity

World→ Area	Electricity		Manufacturing industries and construction		Residential buildings and public services		Transport		Other sectors	
	1990	2014	1990	2014	1990	2014	1990	2014	1990	2014
World	43.0	49.0	20.0	20.0	13.1	8.6	20.0	20.4	3.7	2.0
East Asia and Pacific	38.0	52.4	30.7	27.7	14.3	5.7	12.9	12.3	4.2	1.9
Europe and Central Asia	47.3	48.1	17.8	13.3	14.3	14.1	15.8	22.4	4.9	2.2
Latin America and Caribbean	30.1	35.5	22.4	18.5	9.4	6.3	34.7	36.3	3.3	3.4
Middle East and North Africa	39.7	46.9	21.1	18.4	10.3	8.7	26.1	24.6	2.8	1.3
North America	44.0	45.3	13.0	9.0	11.8	11.4	29.7	33.2	1.4	1.1
South Asia	41.9	52.1	27.8	25.8	11.3	6.1	13.8	13.0	5.2	2.9
Sub-Saharan Africa	50.8	54.8	21.7	13.0	6.4	6.1	18.9	22.9	2.1	3.3
Low income	16.5	22.1	49.0	28.0	3.2	7.5	12.1	32.4	19.2	10.0
Lower middle income	43.4	48.6	25.6	23.1	9.2	7.6	13.9	17.8	7.8	2.9
Upper middle income	44.2	51.1	23.3	25.2	13.2	6.9	14.1	14.6	5.2	2.3
High income	43.2	47.0	16.6	12.8	13.8	10.9	24.7	28.1	1.8	1.2

DI2.1: Syngas and the Fischer–Tropsch process

Syngas (synthesis gas) is a mixture of CO and H_2 in variable molar ratios (usually in the range 1–2.5 H_2/CO) produced by *reforming* fossil-C (coal, methane, hydrocarbons, Eqs. 2.4–2.6) with water, a strongly endothermic reaction that occurs at high temperature, or gasifying (Eq. 2.7) any carbon-based feedstock (including cellulosic and oily biomass or waste fresh organics) under starved oxygen conditions.

$$C + H_2O \rightarrow CO + H_2 \tag{2.4}$$

$$CH_4 + H_2O \rightarrow CO + 3H_2 \tag{2.5}$$

$$CxHy + xH_2O \rightarrow xCO + (y/2 + x)H_2 \tag{2.6}$$

$$Biomass + H_2O\ (O_2) \rightarrow xCO + yH_2 + zCO_2 + vCH_4 + wH_2O \tag{2.7}$$

Equation 2.5 is also said *"wet reforming"* of methane. Equation 2.7 originates *"bio-syngas,"* a complex mixture of gases made mainly of H_2+CO, which may require treatment according to the use it is planned for.

Usually, the reforming is followed by a second reaction (Eq. 2.8) that adjusts the hydrogen content of syngas so to reach the best composition required by the Fischer–Tropsch (FT) process (the most suited ratio for the use of syngas in FT is 2.05 < H_2/CO < 2.15). As shown in Eq. 2.8, part of CO is converted into CO_2 with production of further H_2. Such reaction is called *"water gas shift-WGS."*

$$CO + H_2O \rightarrow CO_2 + H_2 \tag{2.8}$$

CO_2 can be eventually separated from H_2 using several technologies (solid and liquid sorbents or membranes).

Reactions 2.4–2.8 are purely thermal reactions carried out at a temperature of 700–1000 °C depending on the feedstock used. In the case of biomass gasification, catalytic processes have been developed to fasten the conversion.

Syngas can be also used as a fuel (it has 50% energy density of natural gas) or for application in advanced power generation (**DI3.1**). Currently, there are more than 270 plants all around the world with some 690 gasifiers. 74 more plants are under construction all around the world with 238 gasifiers and 83 $MW_{thermal}$. The higher capacity is installed in China. The global market of syngas at 2019 has been estimated around 277 507 $MW_{thermal}$ with a projection to 501 932 $MW_{thermal}$ by 2024, at a CAGR of 12.6% from 2019 to 2024 [7] (1 $MW_{thermal}$ = 3 $MW_{electric}$).

An alternative approach to the production of syngas is the catalytic, strongly endothermic *"Dry Reforming of Methane-DRM"* (Eq. 2.9) in which CO_2 is used as a dehydrogenating agent (oxidant), an attractive option for CO_2 utilization within a circular economy frame.

$$CH_4 + CO_2 \rightarrow 2\ CO + 2H_2 \quad \Delta H_{298} = 248\ kJ\ mol^{-1} \tag{2.9}$$

Notably, DRM converts two major greenhouse gases, CO_2 and CH_4, into useful syngas. The H_2/CO ratio is 1, which is lower than that achieved by steam reforming, and even lower than the ideal ratio for FT (2.05–2.15). By combining reactions 2.5 and 2.8 with reaction 2.9, it is possible to adjust the H_2/CO ratio to the most suited value. The major technological challenges for the industrial application of DRM are its high endothermicity, requiring high temperatures for appreciable conversion, and the deactivation of catalysts under reaction conditions, caused by coking or sintering. The high-energy input required by the process causes a serious penalty if fossil-C has to be used as source of energy and calls for using solar heating to supplement the required energy. For industrial application of DRM, cost-effective catalysts must be developed that can maintain a stable performance for long time. In the past decade, there has been a substantial increase in research focus on catalyst development for DRM and tremendous progress has been made in increasing the activity and stability of catalysts even in a plasma environment [8].

The high endothermicity of the reaction is due to the fact that both CH_4 and CO_2 are very stable molecules with high bond dissociation energy (435 kJ/mol for CH_3–H and 526 kJ mol^{-1} for CO–O), so that high temperatures (>800 °C) are required. Moreover, the H_2/CO ratio is influenced by the simultaneous occurrence of "*Reverse Water Gas Shift-RWGS*" reaction (Eq. 2.10) which further lowers the H_2/CO ratio to <1 by producing water.

$$H_2 + CO_2 \rightarrow H_2O + CO \tag{2.10}$$

Based on the relative endothermicity of the DRM and RWGS reactions, the effect of RWGS on product selectivity is more significant in the temperature range 400–800 °C [9]. Other significant side reactions in DRM are methane decomposition (Eq. 2.11) and the CO disproportionation (Boudouard equilibrium) (Eq. 2.12).

$$CH_4 \rightarrow C + 2H_2 \; \Delta H_{298} = 75 \text{ kJ mol}^{-1} \tag{2.11}$$

$$2CO \rightleftarrows C + CO_2 \tag{2.12}$$

Both reactions 2.11 and 2.12 produce solid carbon (or coke) that can cover the catalyst (coking) and cause rapid deactivation in the temperature range of 557–700 °C (methane decomposition) (Boudouard reaction) [10]. It looks like the optimum temperature at the feed ratio of CO_2/CH_4 = 1:1 is between 870 and 1040 °C, considering the formation and conversion of carbon. Modeling the reaction has confirmed that the optimum working temperature is 850 °C at low pressures for a high conversion [9]. Syngas is largely used in catalytic FT processes for Cn hydrocarbon (Eq. 2.13), olefin (Eq. 2.14), and alcohol (Eq. 2.15) production.

$$nCO + (2n+1)H_2 = C_nH_{2n+2} + nH_2O \tag{2.13}$$

$$nCO + 2nH_2 = C_nH_{2n} + nH_2O \tag{2.14}$$

$$nCO + 2nH_2 = C_nH_{2n+1}OH + (n-1)H_2O \qquad (2.15)$$

Reactions 2.13–2.15 are all exothermic equilibria due to the use of hydrogen as energy carrier. Searching the past for the development of FT, one can identify the following key dates:

- 1902, Sabatier and Senderens discover the synthesis of methane from C and H_2;
- 1913, BASF uses Co–Os catalysts for the synthesis of hydrocarbons from syngas;
- 1923, the first Fischer experiments brought to the synthesis of hydrocarbons, with rich fractions of oxygenates using Fe–K_2O catalysts;
- 1925, Fe–Zn catalysts are advantageously used in FT processes;
- 1936, Ruhrchemie uses Co–ThO_2–MgO, SiO_2 catalysts in the conversion of syngas;
- 1938–1945, the FT process is commercialized in Germany to produce liquid fuels from coal, and since then has received much attention for its high potential of making liquid fuels (gasoline and diesel) from solid (coal) or gaseous (methane) fossil-C at competitive price with those derived from fossil oil.

The variants of the FT process use *catalysts* based mainly on iron (Fe), cobalt (Co), ruthenium (Ru), osmium (Os), and nickel (Ni) (each promoting selectivity toward diverse classes of hydrocarbons) with Group 1 (K_2O, mainly) or ThO_2, CuO, Al_2O_3, Cr_2O_3, TiO_2, MgO as additives/co-catalysts, depending upon the desired product distribution and SiO_2 as support. The process parameters (temperature and pressure) also influence the product distribution [11, 12]. The use of iron-based catalysts is often preferred because of their high activity as well as their participation in the water–gas shift reaction [13, 14].

The reaction mechanism in FT has been extensively investigated since the early years and several pathways have been proposed based on some experimental and spectroscopic evidence. In general, the following steps are considered to play a fundamental role:

   i. Adsorption of CO and H_2 on the catalyst;
  ii. Dissociation of H_2 to 2H;
 iii. Dissociation of CO to C+O;
 iv. Chain initiation;
  v. Chain propagation;
 vi. Chain termination; and
vii. Desorption.

The early hypotheses put forward by Fischer were based on the observation of the formation of abundant oxygenate fractions [15]. In 1926, the *"carbide"* hypothesis was formulated by Fischer–Tropsch as shown in Scheme 2.3 [16].

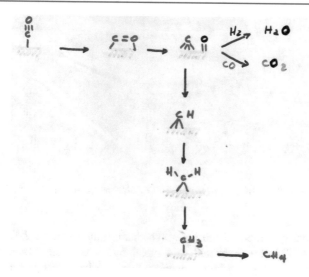

Scheme 2.3 The *"Carbide"* putative mechanism

According to such hypothesis, CO dissociates on the catalyst surface and produces a *carbide* and an *oxide* (Scheme 2.3, upper part). The latter originates water (by reaction with H_2) or CO_2 (by reaction with CO).

The *"carbide"* is stepwise hydrogenated to *carbine* (C–H), *carbene* (CH_2), *methyl moiety* (CH_3), and may end with the formation of methane (CH_4), a not particularly desired product of the FT process. The carbine, carbene, and methyl species may give rise to C–C formation. For example, the carbene "M=CH_2" and methyl "M–CH_3" moieties may give rise to a M–CH_2–CH_3 species and further insertion of a carbene moiety into the M–C bond may grow the chain "M–CH_2– $(CH_2)_n$–CH_3" terminated by either hydrogenation of the M–CH_2–CH_2–CH_2.... bond that affords M–H and H–CH_2–CH_2–CH_2.... moieties or by β-H transfer from the chain to the metal center with formation of a M–H and an *olefin* CH_2=CH– CH_2....moiety (olefins are present in the FT reaction mixture). More sophisticated mechanisms were proposed later on by Craxford and Rideal [17] working with Co catalysts.

Noteworthy, at the very beginning of the FT story, operando-spectroscopic techniques were not available and confirmation of mechanisms was searched through out the study of reactions of model metal organic species under mild conditions. With the advent of surface science, the carbide mechanism was confirmed as abundant carbon was found on the catalyst surface and very scarce oxygen. The role of the *"methylene"* moiety was confirmed by using diazomethane, CH_2N_2, which was a source of –CH_2– which undergo polymerization to long-chain hydrocarbons.

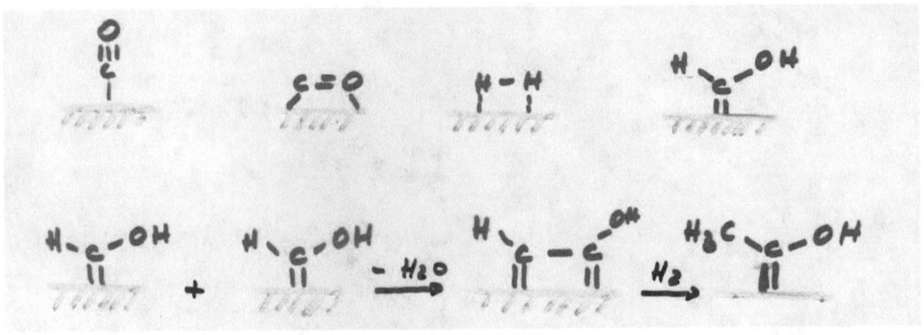

Scheme 2.4 The hypothesis of "*hydroxycarbene =CH(OH)*" as the starter of chain growth

In 1951, Andersen proposed the hydroxycarbene route (Scheme 2.4) in which bound-CO (M–CO) is hydrogenated to a metal hydroxycarbene "M=C(OH)H," which undergoes coupling and hydrogenation. For long time, this hypothesis was silent as there were no examples of stable hydroxycarbene metal organic species. The discovery by Bercaw in 1983 of stable Zr–O–C(H)=Nb species gave new life to the hydroxycarbene route.

The insertion of CO into the M–H bond was also supposed as the start of the reaction that eventually merged into the carbene hypothesis again (Scheme 2.5).

As a matter of fact, different mechanisms may operate at different temperatures and on different metals and this increases the complexity of the reactive system. In fact, when the catalyst used is iron-based, a range of temperatures 300–350 °C are used and this constitutes the High-Temperature (HTFT) process. On the other hand, if the catalyst selected is cobalt-based, the temperature range is 200–240 °C, which represents the Low-Temperature (LTFT) process. But the latter process can also work with iron, it is not specific of Co. The chemistry of the FT process is really vast and complex and there is not a single answer to the question: how syngas is converted into Cn species? The original oxygenate mechanism recently came back again.

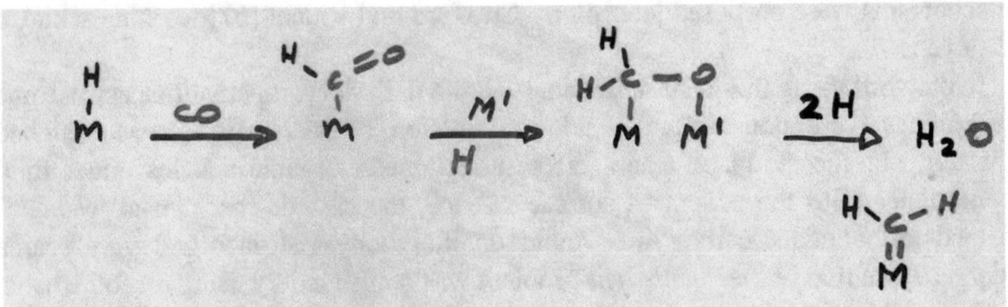

Scheme 2.5 The "*insertion*" hypothesis

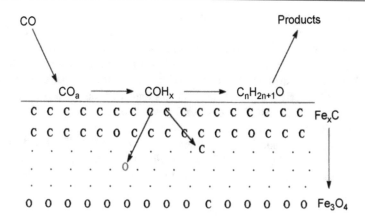

Scheme 2.6 Reaction mechanism in FT under Fe catalysis. Reprinted from Ref. [18], Copyright (2009), with permission from Elsevier

In time, evidence was found for this or that mechanism. The iron catalysts were the most studied because they are the most used in plants and some new hypotheses have been put forward which, for example, separate the nature of the *"initiator"* from that of the *"propagator."* Even a question on the putative activity of iron oxides was risen. Experimental spectroscopic data seem to suggest that the active surface is made of iron carbide, while oxides can be in the internal part of the catalyst particles and do not play a role in C–C bond formation (Scheme 2.6).

In modern plants, the gasifier is integrated with the FT reactor, and the overall design is aimed at achieving a high yield of liquid hydrocarbon products. The formation of methane and other gaseous product, as well as the formation of solid waxes, must be avoided as much as possible, and the production of gasoline and diesel maximized, for a higher economic value [19]. Nowadays, a lot of progresses have been made avoiding the production of waste in FT: gaseous products are valorized via C–C bond formation to produce higher hydrocarbons, while waxes are hydrogenolysed to produce shorter chains usable as fuels.

As it is clear from the above discussion, the FT reactive system is very complex as it contains reagents such as CO, CO_2, H_2O, H_2, and CH_4 which can give rise to multiple equilibria. Accordingly, the optimization of conversion of syngas into most valuable products (gasoline and diesel) requires a great effort in terms of catalyst modification (by changing the catalyst, the promoters, the acid and basic components, etc.) and formulation. After 100 years still a great effort is put in optimization of the process (catalysts and reactors) because FT process is an interesting route for coal-to-liquid conversion. Notably, liquid fuels obtained by the FT process are much cleaner than oil-derived gasoline and diesel as they do not contain sulfur, which is present in fossil-C, mainly in coal. Their cost per barrel may vary with the location where they are produced. FT processes are commercial on a large scale in Malaysia and South Africa where gasoline and diesel for local use are made from coal by combining syngas–FT processes, but are also present all over the world. Anyway, even if the production cost of gasoline and diesel is competitive with the price of production from fossil oil (even at a cost of the latter of 60 US$/barrel), the main barrier to FT wide spreading is the CAPEX costs. In fact, it takes only a few weeks to drill a well using fracking technology with an investment near 1 MUS$, and the well can pay for itself in less than 12 months, while a FT facility can cost up to hundreds MUS$, take 2 or 3 years to bring into production, and then take several years to pay for itself.

Nevertheless, under specific circumstances (national abundance of coal and scarce access to oil and gas) the syngas–FT combination has shown a great role for making liquid fuels from coal and feed the transport sector (terrestrial, avio, and maritime).

References

1. Total primary energy supply 1990–2017. https://www.iea.org/data-and-statistics
2. Aresta M, Dibenedetto A, Nocito F (2018) What catalysis can do for boosting CO_2 conversion. Adv Catal 62:49–111
3. Aresta M, Karimi I, Kawi S (eds) (2019) An economy based on carbon dioxide and water. Springer
4. Major environmental aspects of gasification-based power generation technologies. Final report. https://www.netl.doe.gov/sites/default/files/netl-file/final-env.pdf
5. https://www.census.gov/library/visualizations/2011/demo/world-population–1950-2050.html
6. https://commons.wikimedia.org/wiki/File:Global_annual_CO2_emissions_by_world_region_since_1750.svg
7. https://www.marketsandmarkets.com/PressReleases/syngas.asp
8. Aresta M, Dibenedetto A, Quaranta E (2016) Reaction mechanisms in carbon dioxide conversion. Springer
9. Pakhare D, Spivey J (2014) A review of dry (CO_2) reforming of methane over noble metal catalysts. Chem Soc Rev 43:7813–7837
10. Wang S, Lu GQ, Millar GJ (1996) Carbon dioxide reforming of methane to produce synthesis gas over metal-supported catalysts: state of the art. Energy Fuels 10:896–904
11. Overend RP (2004) Thermochemical conversion of biomass. In: Shpilrain EE (ed) Renewable energy sources charged with energy from the sun and originated from earth-moon interaction. EOLSS Publishers, Oxford, UK
12. Chadeesingh R (2011) The Fischer–Tropsch process. In: Speight JG (ed) The biofuels handbook. Part 3. The Royal Society of Chemistry, London, UK. Chapter 5, pp 476–517
13. Rao VUS, Stiegel GJ, Cinquegrane GJ, Srivastava RD (1992) Iron-based catalysts for slurry-phase Fischer Trospch process: technology review. Fuel Proc Technol 30:83–107
14. Jothimurugesan K, Goodwin JG, Santosh SK, Spivey JJ (2000) Development of Fe Fischer–Tropsch catalysts for slurry bubble column reactors. Catal Today 58:335–344
15. Fischer F (1925) The conversion of coal into oils. In: Lessing R (trans). Ernest Benn Ltd., London
16. Fischer F, Tropsch H (1926) German patent 484, 337, 195. The synthesis of Petroleum at atmospheric pressure from gasification products of coal. Brennstoff-Chem 7:97–104
17. Craxford SR, Rideal EK (1939) The mechanism of the synthesis of hydrocarbons from water gas. J Chem Soc 1604–1617

18. Davis BH (2009) Fischer–Tropsch synthesis: reaction mechanisms for iron catalysts. Catal Today 141:25–33
19. Balat M (2006) Sustainable transportation fuels from biomass materials. Energy Edu Sci Technol 17:83–103

The Atmosphere, the Natural Cycles, and the "Greenhouse Effect"

3

Abstract

In this chapter, the structure of the atmosphere is presented with the natural cycles of several species. The natural green house effect is discussed and the way in which CO_2 and other GHGs contribute to its enhancement is analyzed.

3.1 The Atmosphere and Its Structure

Our planet Earth is surrounded by a gaseous mass that is structured as shown in Fig. 3.1.

The troposphere is the layer closer to the surface of our planet. The composition of the atmosphere is given in Chap. 1, W1. It varies with altitude, as heavier gases are more abundant close to the surface and lighter gases in the higher layers. The estimated total mass of the atmosphere is $5.14 \ 10^{15}$ t with roughly 80% in the troposphere, [1b] its average density 1.2 g/L, and the average mass of water present in it $1.25 \ 10^{13}$ t mainly as vapor or condensed forms at high altitude. The temperature of the atmosphere decreases quite rapidly from 0 to 13 km above the surface where it can be as low as −60 °C, then remains quite constant in the tropopause and starts to grow again up to *ca.* 0 °C at 50 km above the surface, and decreases again up to −80 °C reaching the height of 85 km. Other parameters of interest for understanding the behavior of the atmosphere are the specific heat at constant pressure (C_p) and at constant volume (C_v) that are equal to 1.00 kJ/kg K and 0.718 kJ/kg K at 27 °C (300 K), respectively. C_p and C_v indicate the amount of heat necessary to rise by 1° the temperature of 1 kg of air at 27 °C (300 K), either keeping constant the pressure (C_p) or the volume of the mass of gas (C_v).

© Springer Nature Switzerland AG 2021
M. Aresta and A. Dibenedetto, *The Carbon Dioxide Revolution*,
https://doi.org/10.1007/978-3-030-59061-1_3

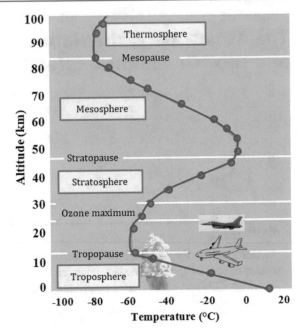

Fig. 3.1 The structure of the atmosphere that envelops our planet [1a]

3.2   The Natural Cycles of Selected Species (Elements and Compounds)

3.2.1   The Carbon Cycle

Whatever C-based good we use, it will be converted into CO_2 at the end of its life and will reach the atmosphere starting its natural "cycle": the so-called "Carbon Cycle" represented in Fig. 3.2.

The natural C-cycle has been developed on Earth in billion years. There is evidence of early autotrophic C-fixation at 3.7–3.8 billion years ago, and first photosynthetic organisms were present some 3.2–3.5 billion years ago on our planet [2b]. The accumulated evidence suggests that (i) photosynthesis began early in Earth's history, but was probably not one of the earliest metabolisms; (ii) the earliest forms of photosynthesis were anoxygenic, with oxygenic forms arising significantly later. However, the C-cycle represents the *"equilibrium of CO_2 uptake – CO_2 release"* that controls the CO_2 atmospheric concentration keeping it at a constant level that was around 270 ppm (or 0.0270% v/v) until the start of the *"Industrial Age."* The *Industrial Age* is a period of history that encompasses the changes in economic and social organization that began around 1760 in Great Britain. From 1830 and later, the industrial age heavily started in Europe and in the geographic area today known as USA, and after in the entire Earth planet. Industry is energivorous and the use of any kind of fuel (wood at the start, and then coal, oil,

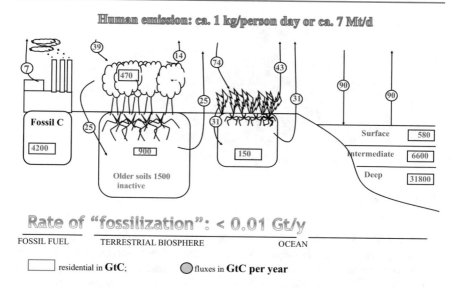

Fig. 3.2 The natural C-cycle. All figures are in Gt of C. Circles represent "fluxes of C," rectangles give "residential-C." Deforestation is causing the reduction of C-fixed as biomass. Human biogenic emission (respiration) is growing due to the increase of population. Adapted from Ref. [2a], Copyright © 2003 American Chemical Society

gas, after the industrial use of such fuels was developed) became more and more intensive, shifting in time from low-density energy sources (wood) to more intensive ones (coal, oil, gas, man-made fuels). With energy becoming available in many forms, even the style of life of mankind improved. The utilization of electricity started a new era. Electricity was known to Greeks since 600 bC, but it was the Italian Alessandro Volta in 1801 who discovered that it is possible to generate steady electric flows and to produce portable-electric energy (electrochemical cell that is the basic unit of batteries). Such discovery has caused a great change in human life.

All chemical species (elements or compounds, except noble gases) present in the atmosphere are subject to cyclic changes in their nature that can be represented in the so-called *"natural cycle."* Some of the most relevant to the topic of this book are represented below.

3.2.2 The Water Cycle

Figure 3.3 shows the *water natural cycle*, which is based on several physical phenomena. Water (H_2O) vapor present in the atmosphere reaches the Earth surface by *precipitation* in various condensed phases (rain, snow, ice) depending on the atmosphere temperature. Some of it *runs off* on the surface (rivers) and reaches the

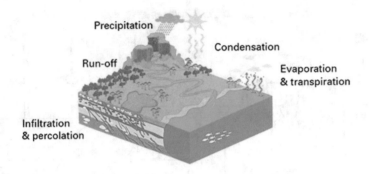

Fig. 3.3 The natural *water cycle* [3]

oceans or lakes; some goes to subsurface by *infiltration or percolation* through soil and rocks; and some reaches again the atmosphere by *evaporation*. It is a complex equilibrium and the intensity and frequency of precipitation phenomena are driven by both the amount of water vapor in the atmosphere and the temperature.

Increasing the temperature increases the evaporation, the accumulation in the atmosphere, and the violence of the precipitation, which is something we are experiencing these days: extreme phenomena can originate by climate change and atmosphere warming or cooling.

3.2.3 The Methane Cycle

The *methane (CH₄) cycle* begins in soil where it is produced by microbes and is consumed by microorganisms that live on it, the so-called *Methanotrophs*. Other microorganisms, said *Methanogens*, produce more methane that *Methanotrophs* consume. Methane is also generated by other sources, such as landfills (anaerobic digestion of fresh organic waste), livestock, and exploitation of fossil fuels. Wetlands are responsible for some 80% of the methane released to the atmosphere. The use of LNG as fuel in cars is increasing the release to the atmosphere due to unburned fuel and adventitious leakage of the tanks. In the troposphere, methane undergoes a series of reactions initiated by the naturally occurring OH⁻ radical that extracts a hydrogen atom and generates the reactive methyl radical, CH₃⁻, which continues loosing hydrogen radicals and finally reacts with atmospheric oxygen: the fate of such carbon is the formation of CO_2.

3.2.4 The Ozone Cycle

Ozone (O_3) cycles between the troposphere (0–10 km) and the stratosphere (up to 45 km) (see Fig. 3.1). It is formed by the splitting of the oxygen molecule into two oxygen atoms. The atoms can react with molecular oxygen and afford ozone,

$$O_2 \rightarrow 2\,O$$

$$O + O_2 \rightarrow O_3$$

$$O_3 \overset{hv}{\rightarrow} O_2 + O$$

Scheme 3.1 Formation and decomposition of ozone

which, in turn, is split by solar UV radiation into atomic and molecular oxygen (Scheme 3.1).

Ozone is essential for "filtering" the solar UV radiations and preventing the radiations from reaching the Earth's surface where it can damage humans, animals, and vegetation.

3.2.5 The NO Cycle

NO, nitrogen oxide, is formed from atmospheric dinitrogen (N_2) in air-based combustion processes. It enters the N-cycle in which it is either oxidized to nitrate (upper right part of Fig. 3.4) and reaches the surface as "HNO_3", component of acid rains, or reduced to N_2 (lower part of Fig. 3.4).

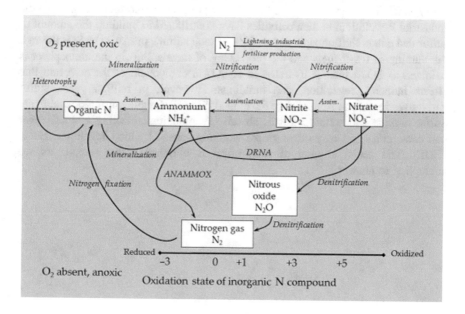

Fig. 3.4 The nitrogen cycle Reprinted from Ref. [4], Copyright (2020), with permission from Elsevier

NO is toxic and its oxidized forms (NO_2, NO_2^-, NO_3^-) are oxidants that can affect living organisms and manufacts. Ammonia is a toxic agent. Therefore, the atmospheric concentration of all N-derivatives must be kept under control.

3.2.6   The Sulfur Cycle

The sulfur (S) cycle encompasses both atmospheric and land processes. On the Earth surface, the sulfur cycle (Fig. 3.5) starts with the weathering of rocks during which the stored sulfur comes into contact with atmospheric oxygen and is eventually converted into sulfate ($SO_4^=$).

The sulfate is uptaken by plants and microorganisms and is converted into organic forms, and enters the animal food chain. When organisms die and decompose, some of the sulfur is again released mainly in the form of sulfate and some enters the tissues of microorganisms.

There are also a variety of natural sources that emit sulfur directly into the atmosphere, including volcanic eruptions, the breakdown of organic matter in swamps and tidal flats, and the evaporation of water. Sulfur eventually settles on the Earth surface or comes down within rainfall reaching oceans. Surface water freely and continuously drains off sulfur from terrestrial ecosystem into lakes and streams, and eventually seas where aquatic communities move sulfur through the food chain.

Sea spray is a direct vector of sulfur back to the atmosphere. Part of sulfur in oceans reaches the ocean depths, combines with iron, and forms ferrous sulfide (FeS) which is responsible for the black color of some marine sediments. Since the Industrial Revolution, human activities have contributed to build up the amount of sulfur that enters the atmosphere (*ca.* 35% of total sulfur), primarily as SO_2 through the burning of fossil fuels and the processing of metals. SO_2 in the atmosphere is oxidized to SO_3 which is hydrated to H_2SO_4, a component of the acid rain that affects humans, vegetation, and buildings. However, in million years, sulfur derivatives in the atmosphere have developed the role of regulator of global climate. Sulfur dioxide and sulfate aerosols absorb UV radiations, creating a cloud cover that cools cities and may offset global warming caused by the Greenhouse Effect (GHE). The actual amount of such offset is a question that researchers are attempting to answer.

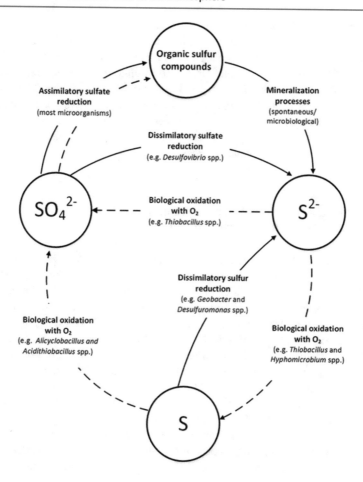

Fig. 3.5 The sulfur cycle Reproduced from Ref. [5] Open access (CC BY 4.0)

3.3 The Chlorofluorocarbons' Fate in the Atmosphere

Chlorofluorocarbons (CFCs) have a general formula $C_nF_xCl_{2n+2-x}$ and were used in aerosols, refrigerators (as substitutes of ammonia (NH_3), sulfur trioxide (SO_3), and methylchloride (CH_3Cl)), air conditioners, foam food packaging, and fire extinguishers. They were synthesized in 1928 with the express target of substituting the abovementioned gases in refrigerators after a huge problem caused by a leakage of toxic methyl chloride. They had a great acceptance and enormous consensus by scientists and the public because they are odorless and non-reactive at the point that before 1996, they were even used against asthma. CFCs were produced at the level of Mt/y in the 1980s–1990s. CFCs are not decomposed in the troposphere and reach the stratosphere unaltered, located 25–30 km above the Earth surface, where they react under UV radiations and generate chlorine radicals, Cl·, that may affect and destroy the ozone layer: one chlorine radical can destroy up to 100 000 molecules of ozone. When their effect was recognized in 1974 at the University of California by Prof. F. Sherwood Rowland and Dr. Mario Molina, serious measures toward the limitation of their use were taken in 1987 with the Montreal Protocol and, finally, most of them were banned on January 1, 1996 and progressively substituted with less stable molecules that are more easily destroyed and do not reach the stratosphere. As an effect of such policy, the "*ozone hole*" strongly decreased over the last 20 years. CFCs also behave as greenhouse gases.

3.4   Greenhouse Gases, Their Atmospheric Level, and Effect on Climate Change

The natural atmospheric concentration of greenhouse gases (GHGs: water vapor—the most abundant, carbon dioxide, methane, nitrous oxide, ozone, and some synthetic chemicals such as CFCs) has been kept under control by the respective natural cycles developed in million years. In Fig. 2.1, the carbon cycle is shown. The continuous increase of the atmospheric level of GHGs and CO_2, in particular, due to anthropogenic activities and the fact that the excess amount cannot be buffered by the natural cycle is rising serious worries about the impact on the environment. Even, such increase is correlated to the climate change on our planet. As mentioned above, the production of energy from fossil-C is performed with low efficiency that averages 32%. This is true for the production of electric energy from coal, oil, and methane, for the production of heat, and for the use of gasoline and diesel in the transport sector. Noteworthy, the remaining 68% thermal energy is released to the atmosphere in the form of high-temperature heat that causes a direct heating of the atmosphere. More fossil-C we burn, more heat we transfer to the atmosphere.

3.5   The Greenhouse Effect

But what is the greenhouse effect? Is it natural? How anthropic activities can influence the natural effect?

Solar radiations (constituted mainly of *visible light*, with *ca.* 8% UV component) during daytime reach the Earth's atmosphere: part is reflected back into the space, part is absorbed by land and oceans, heating the Earth (Fig. 3.6). During night, heat radiates from Earth toward the space in the form of *infrared radiations*: some of this heat is trapped by *greenhouse gases* present in the atmosphere, keeping the Earth warm enough (average temperature 13–14 °C) to sustain life. Without the "*greenhouse effect (GHE)*" the Earth average temperature would be some 33 °C lower making impossible life to exist. However, the *greenhouse effect* is natural and

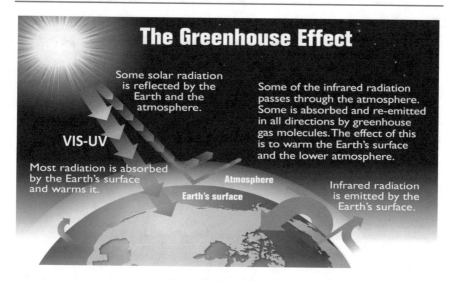

Fig. 3.6 The greenhouse effect: how it is generated and its role in controlling the temperature of our planet [6]

we could not live without it (Fig. 3.6). GHGs, as mentioned above, include water vapor (the most abundant), carbon dioxide, methane, nitrous oxide, ozone, and some synthetic chemicals such as CFCs.

To act as GHG, a species must absorb infrared radiations (Fig. 3.7) emitted from the Earth's surface. *Infrared radiations* extend from the nominal red edge of the visible spectrum at 700 nm to 1 mm. This range of wavelengths corresponds to a frequency range of approximately 430 THz down to 300 GHz. Figure 3.7 shows the absorption bands of H_2O and CO_2 at the various wavelengths.

It is quite evident that H_2O has a much larger absorption power than CO_2, which absorbs in only four regions, namely, at 2.0, 2.7, 4.3, and 15 μm. CO_2, on the other hand, reinforces the water absorption power in the range 13–18 μm. There is still an argument about CO_2 being responsible of climate change [8], with some scientists saying that even doubling the atmospheric CO_2 concentration would not cause any climate change (see Chap. 12).

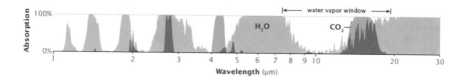

Fig. 3.7 Absorption spectra in the IR region of H_2O and CO_2 [7]

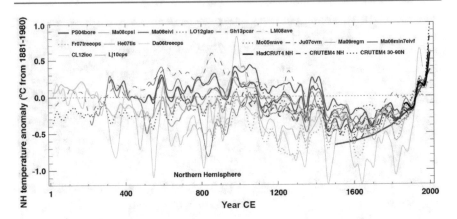

Fig. 3.8 Temperature trend in the Northern hemisphere over the last 2000 years [9]

As said above, the increased atmospheric concentration of CO_2 witnesses that high temperature heat is being transferred to the atmosphere with direct increased heating. The increased temperature causes faster water evaporation and accumulation of water vapor in the atmosphere. From one side, this will cause a larger absorption of IR radiations and from the other will cause heavy downpours. Violent rains are not beneficial to soil. In fact, with a swift rain, water has time to permeate soil and reach subsurface reservoirs with a double benefit of increasing the humidity of deep layers of soil and enrich soft water reserve. With violent rains, water runs away even causing damages.

However, the GHE is naturally caused by GHGs present in the atmosphere: if the GHG concentration is increased the escape of radiations is slowed down with increase of the heat trapped in the atmosphere and transferred to oceans and soil. This causes climatic changes and extreme events.

It must be said that temperature waving is also a natural phenomenon as shown in Fig. 3.8.

Periodic ups and downs of temperature represent, thus, the normality: what is worrying these days is the continuous anomalous increase of the average temperature from the start of the industrial age (right part of Fig. 3.8). Such increase is being correlated to the increase of atmospheric concentration of GHGs and of CO_2, in particular. But is CO_2 the real responsible for such increase? Such question demands further studies and debate prior a clear answer is given. In the meanwhile, the responsible care principle imposes that measures are taken for reducing its emission, even making less use of fossil-C, saving it for next generations.

3.6 The CO$_2$ Accumulation in the Atmosphere and Correlation to Health of Our Planet

The increase of the atmospheric level of CO$_2$ has been correlated to the climatic changes observed in recent years. Figure 3.9 [10] shows a stable situation between 1850 and 1920, a continuous increase since 1920 with two interruptions during 1945–1975 and 1998–2012.

Since 2011 the temperature started to rise again. 1990 is taken as zero in Fig. 3.9. Between the two pauses the temperature growth was of 1 °C. Since 1990 we had an increase of 0.9–1.0 °C.

The worrying issue is that a rise of the ocean surface has been detected (Fig. 3.10). It originates from a number of sources. Melting of ices in the polar regions and at high quote in mountains makes that "*stable on land water*" (ice) is now flowing to the oceans. Moreover, the increase of oceans' temperature is causing their expansion. The temperature of seawater is measured at various deepness using a variety of techniques and means, including research boats. An increase of the temperature of oceans makes that the volume of the ocean is expanding as the water warms. A third much smaller contributor to sea level rise is a decline in the amount of "liquid water on land": aquifers, lakes and reservoirs, rivers, and soil moisture. This shift of liquid water from land to ocean is largely due to groundwater pumping.

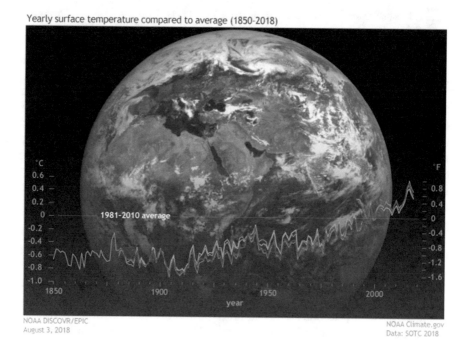

Fig. 3.9 Trend of earth surface average temperature since 1850 [10]

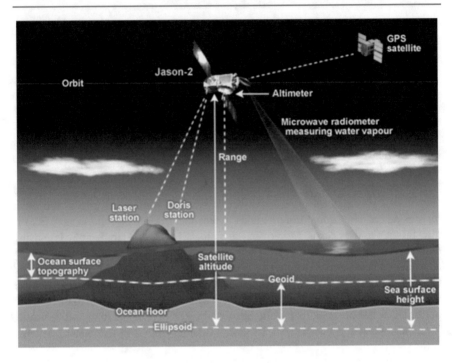

Fig. 3.10 Satellite technology for measuring the sea level [11]

The latter practice also increases the infiltration of salt water in aquifers, lowering the quality of such water and its use in agriculture and as drinking water.

However, the global average water level in the oceans rose by 3.6 mm/y from 2006 to 2015, which was 2.5 times the average rate of 1.4 mm/y throughout most of the twentieth century. Such trend brings to a rise, by the end of this century, of 0.3 m above 2000s' levels, even wishing that GHG emission follows a relatively low trend during next 80 years. Heat is stored at different rates in the oceans, depending on deepness. Even winds and marine currents make a difference and can strongly influence tides. Therefore, the rise of the water level may vary with the geographical area and threatens several coastal communities. Figure 3.10 shows the sophisticated tools used today for measuring the sea level [11].

Until the 1990s, it has been challenging to measure the average sea level because the ocean's surface is not flat, it changes daily or hourly based on winds, tides, and currents. The old technique was based on tide gages that were placed on piers and continuously recorded the height of the water level compared to a stable reference point on land. Roughly, 2000 tide gages around the world, run by around 200 countries, gave the possibility of determining the average sea level. Some tide gages have been recording sea level data since the 1800s and a few for even longer.

In 1992, NASA launched TOPEX/Poseidon, the first of a series of satellites that measure sea level rise from space. It was followed by Jason-1 and OSTM/Jason-2, and most recently Jason-3 which was launched successfully on January 17, 2016. Such satellites allow a good knowledge of the status of the oceans and give precious information on local anomalies.

References

1. (a) https://climate.ncsu.edu/edu/Structure; (b) http://acmg.seas.harvard.edu/people/faculty/djj/book/bookchap2.html
2. (a) Aresta M (2003) Carbon dioxide utilization: greening both the energy and chemical industry: an overview. In: Liu C-J, Mallison RG, Aresta M (eds) Utilization of greenhouse gases. ACS symposium series, vol 852, pp 2–39; (b) Cardona T (2018) Early Archean origin of heterodimeric Photosystem I. Heliyon 4(3):e00548
3. https://www.sydneywater.com.au/SW/education/drinking-water/Natural-water-cycle/index.htm
4. Dodds WK, Whiles MR (2020) Nitrogen, sulfur, phosphorus, and other nutrients. Freshwater ecology, 3rd edn. Aquatic ecology, pp 395–424
5. Huber B, Herzog B, Drewes JE, Koch K, Muller E (2016) Characterization of sulfur oxidizing bacteria related to biogenic sulfuric acid corrosion in sludge digesters. BMC Microbiol 16:153
6. https://commons.wikimedia.org/wiki/File:Earth%27s_greenhouse_effect_(US_EPA,_2012).png
7. http://climateilluminated.com/climate/climate_science/26_physics_IR_absorption.html
8. https://nov79.com/gbwm/ntyg.html
9. https://www.ncdc.noaa.gov/global-warming/last-2000-years
10. https://www.climate.gov/news-features/understanding-climate/climate-change-global-temperature
11. https://theconversation.com/sea-level-is-rising-fast-and-it-seems-to-be-speeding-up-39253

Reduction of the CO_2 Production

4

Abstract

The efficiency technologies in both the production of energy (electric–thermal–mechanic, others) from fossil-C and the use of the various forms of derived energies can play a strategic role in the reduction of the production and emission of CO_2. Additionally, the use of biomass as source of fuels is shortly presented and the putative *"zero-emission"* property discussed.

4.1 How Much of the GHGs Emission Should Be Cut? And by When?

The CO_2 balance of a process/product/service is *the Carbon Footprint (CF)*. The *CF* reduction can be targeted using different strategies that can be summarized in three intervention lines:

- Implementation of efficiency strategies e.g. in the production and use of energy.
- Substitution of fossil-C with alternative sources that emit less CO_2.
- Substitution of fossil-C with alternative sources that do not emit CO_2.

The latter approach will be discussed in detail in Chap. 5. In this chapter, we deal with the efficiency strategies both in the production and use of energy and with the use of biomass.

Let us start with the specification of how much CO_2 should be avoided. This is an important target to be set in order to design routes to reach it.

Figure 4.1 shows that over the time the concentration of atmospheric CO_2 has always had highs and lows. In particular, low peaks correspond to low temperatures (glaciations). The continuous recent increase of the atmospheric CO_2 level has started with the industrial age (end of 1700) when the massive use of fossil-C

© Springer Nature Switzerland AG 2021
M. Aresta and A. Dibenedetto, *The Carbon Dioxide Revolution*,
https://doi.org/10.1007/978-3-030-59061-1_4

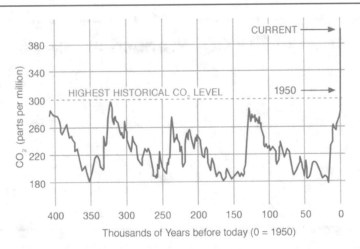

Fig. 4.1 Trend of atmospheric CO_2 concentration [1]

started. As a matter of fact, the atmospheric level of CO_2 was 273 ppm before industrial age and has reached 410 ppm in 2019. Several international conferences have deliberated rules to follow for controlling the CO_2 emission. For example, already in 1992, a great concern about the impact on climate of anthropic activities was expressed at the level of United Nations Organizations, where the *equity principle* and *differential responsibilities* were debated.

The resolution of UN-FCCC sounds as follows:

> *protect the climate system for the benefit of present and future generations of humankind, on the basis of equity and in accordance with their common but differentiated responsibilities and respective capabilities.* (UNFCCC, 1992: Art 3)

Developed countries should, thus, make a wise use of resources using advanced technologies, favoring the development of less advanced societies. Has such recommendation been implemented internationally? Let us see where we are.

Differently from what happened with CFCs, when the agreed reduction and banning was implemented worldwide, with the CO_2 reduction several countries have adopted an ambiguous attitude toward the implementation of agreed policies. The 2015 COP 21 Paris Climate Conference has produced resolutions which go in the direction of a reduced use of fossil-C and sustain to innovation with the target of not trespassing the limit of 2° increase of the planet average temperature [2]. Nevertheless, major players such as USA have first signed the document and then withdrawn from its implementation due to the change in political and economic views. At the international level, several agreements have been signed which target the CO_2 emission reduction. For example, the European Union (EU) has agreed the so-called "202020" and the "502050" agreements. The former states a reduction of the emission of CO_2 by 20% within 2020 with respect to the 1990 emission and the latter a reduction of 50% by 2050. The first goal was not fully reached at the pan-EU level, countries are strongly engaged with the adoption of several different policies to reach the second. As mentioned above, the reduction of the emission is finalized to stabilize the atmospheric CO_2 level. The CO_2 emission global budget for limiting global warming to 2 °C was calculated in 2013 by the Intergovernmental Panel on Climate Change (IPCC): to reach the target it would be necessary to stabilize the CO_2 atmospheric level at 450 ppm with a 50% likelihood of success [2b]. During 2019 the atmospheric concentration of CO_2 has already reached 410 ppm and continues to growth.

The challenge facing the world to limit future warming to tolerable levels is extraordinarily fearful, and will likely require a level of global cooperation far beyond any other previous environmental problem. Differently from the case of CFCs for which substitutes were relatively easy to find, substituting fossil-C for preventing CO_2 emission is not an easy task. Fossil-C touches all levels or our life, everywhere. Fossil fuels have the highest energy density with respect to other options (vide infra) and can be easily transported everywhere and used. Some countries have large reserves of fossil-C and can live for decades without any threat of energy scarcity. Therefore, they have the tendency to stay on the border of the problem: they will not be saved in case of extreme events that are global. There is really a need of full international cooperation to problem-solving. However, once the atmospheric stabilization goal is fixed at 450 ppm, it is possible to calculate the budget of total GHGs emissions for the *entire world* in order to achieve that atmospheric stabilization goal. Such global budget obliges national, state, and regional authorities to implement GHG emissions reduction targets and, consequently, it drives the energy–industry–transport-standard of life policies worldwide. If some countries do not implement the correct policies, others should implement stricter measures to keep the budget constant, which is not in the direction of the UN-FCCC resolution. Nations should not ignore their ethical duties to the rest of the world and should adopt energy policies for the implementation of equity principles and responsible care. The observance of rules is frequently an explicit element in international discussions about climate change policies and among countries still today exist a variety of different behaviors [3].

But let us analyze some figures about the "2 °C *maximum rise of temperature.*"

To have an approximately 66% chance ([2a]) of keeping warming of our planet below 2 °C, our global society must work as a single community to avoid that global GHG emissions may exceed *ca.* 270 Gt_{CO2eq}. It is worth recalling that we are considering the bunch of GHGs and not only CO_2; for this reason, the unit Gt_{CO2eq} is used, which means that the environmental effect of a unit of each GHG is expressed in terms of an equivalent amount of CO_2 that would produce the same effect. Table 4.1 reports the GHGs as listed at the Kyoto 2007 IPCC Meeting.

The "global warming potential-GWP" of a GHG, reported in Table 4.1, indicates the amount of warming a gas causes over a given period of time (normally 100 years). GWP is an index, with CO_2 having the reference index value of 1, and the GWP for all other GHGs is the number of times more warming they cause compared to CO_2. For example, 1 kg of methane causes 25 times more warming over a 100-year period compared to 1 kg of CO_2, and so methane has a GWP of 25.

Table 4.1 Kyoto greenhouse gases (listed at the Kyoto-IPCC 2007)

Entry	Species	Global Warming Potential (GWP)
1	**Carbon dioxide (CO_2)**	**1**
2	Methane (CH_4)	25
3	**Nitrous oxide (N_2O)**	**298**
4	Hydrofluorocarbons (HFCs)	124–14 800
5	**Perfluorocarbons (PFCs)**	**7 390–12 200**
6	Sulfur hexafluoride (SF_6)	22 800
7	**Nitrogen trifluoride (NF_3)**	**17 200**

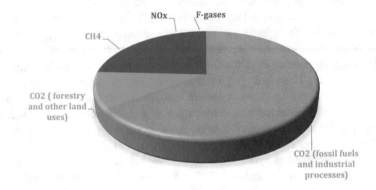

Fig. 4.2 Emission of various GHGs

However, all GHGs (Fig. 4.2) must be taken under strict control if we wish to obey the mandate of not going beyond the established limit of global emission.

Such amount can be emitted over a variable scale of time: when the limit of 270 Gt_{CO2eq} has been reached, the entire world's GHGs emissions should fall to zero to give reasonable hope of limiting warming to 2 °C. This is a dramatic condition and all countries should elaborate a national policy for meeting the common target.

However, since the world is now emitting carbon dioxide equivalent emissions at the approximate rate of 30 Gt/y, the world will run out of emissions under the global budget in approximately 10 years. This is really worrying and demands for immediate action, if the impact of GHGs on climate change is in reality what is believed.

A way to reduce the emission of CO_2 is increasing the efficiency of the conversion of chemical energy of the fossil fuels into other forms of energy. We shall consider below the cases of production of electric energy and the transport sector, which are responsible for the emission of more than 60% of total CO_2.

4.2 Efficiency in the Production of Electric Energy

Figure 4.3 compares the "average" efficiency in the conversion of several primary sources into electric energy. Some data do not consider specific applications of technologies which are proved at the demo scale, but are not yet spread in the real world. As an example, the efficiency of generation of electric energy from coal varies widely with the technology used (**see DI3.1 for a more detailed discussion**). In a traditional (old style and still most used technology throughout the world) coal plant, for example, only about 28–33% of the chemical energy ends up as electricity, which has been improved by 20 points since 1985 through higher steam operating temperatures and heat recovery.

"*Supercritical*" coal plants can reach efficiency levels around 40–46%, and the most innovative integrated gasification combined cycle or IGCC (only a few plants worldwide) reaches efficiency above 60%. The most efficient gas-fired power plants achieve a similar level of efficiency. Lignite-fired power plants may achieve 43% efficiency once the issue of drying of lignite is set.

Doubling the efficiency would mean halving the CO_2 emission!

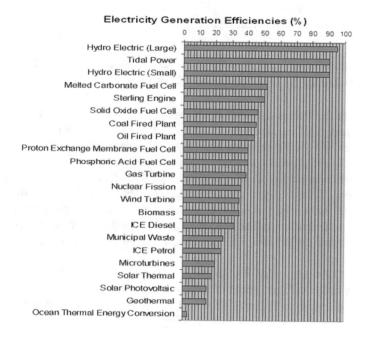

Fig. 4.3 Efficiency in the production of electric energy. Reproduced from Ref. [4a] licence CC BY 4.0

Noteworthy, efficiency is not an issue in the use of perennial primary energies (SWGH) (see Chap. 5) for the simple reason that "fuels" like wind and sunlight do not emit CO$_2$ and the environmental impact on climate change of their use is not a real issue. Moreover, they are free and easily available.

Limiting the analysis of the efficiency only to the production of electricity from fossil-C does not tell the whole story. In fact, the first negative impact is the poor heat to steam conversion (*ca.* 41–46% depending on the technology) while the electricity production by alternators ranges around 98–99%. The overall efficiency is lowered to 32–35% with the various necessary cleaning technologies linked to electricity production (abatement of SOx or NOy). This amount of electric energy is not yet what consumers use, for example, in houses. In fact, transmission of the electrical energy over the distribution grid between the power station and the consumer results in a distribution loss of *ca.* 10% mainly due to the resistance and dispersion of the electrical cables. Further energy is lost in voltage conversion at the end user's appliance. Additionally, incandescent lighting is particularly inefficient so that finally only 5–10% of the electrical energy is converted into light when using old tungsten lamps, which are now out of use. Efficiency can be increased along the entire production–utilization chain. For example, modern lamps are six to seven times more efficient than the old tungsten ones [4b], but new tungsten bulbs are coming back with excellent performances.

4.3   Efficiency in the Transport Sector

Considering the transport sector, in an average car only 20–22% of the original chemical energy of the fuel is used for moving the vehicle, the rest is lost in the engine itself (61–63%), standing idle (16–17%), drive train (5–7%), and various parts (1–2%). Diesel cars are more performant than gasoline cars due to the use of direct injection in the former category. But such difference is going to disappear in coming years. Turbocharged direct injection engines in the European market are estimated to attain 18% reduced fuel consumption, part of which is due to intake valve control and other engine technologies. Other important technologies include cylinder shutoff during low load conditions and improved valve timing and lift control [5].

Great efforts are made for improving the efficiency of energy conversion worldwide in the various sectors of transport. Good results are obtained under the push of legislation in EU and USA and other countries. Most advanced truck engines can reach an efficiency of 45% today.

The EU legislation has put strict limitations to emissions by cars in terms of NOx and CO$_2$ per km with the various Euro 4, 5, 6 standards. The latter tends to limit the CO$_2$ emission to less than 100 g/km. Today, in many cities, the Euro 4 cars are not admitted. The introduction of "hybrid cars" is further improving the efficiency of fuels as such cars can save energy in various ways, such as:

- Shutting the engine down when the vehicle is stopped, leaving on the electric motor;
- Using the electric motor to brake and using the electricity generated to recharge the battery;
- Using the electric motor to start or boost power during acceleration;
- Using the electric motor instead of the engine at low speed and load (that saves emissions in city circuits);
- Allowing the use of a more efficient cycle than the standard Otto cycle (in some hybrids); and
- Shifting power steering and other accessories to (more efficient) electric operation.

We do not consider here the "full electric cars" as they do not rely on liquid fuels as power source, but receive energy from outside and represent a different case.

4.4   Energy Utilization Efficiency

Besides increasing the efficiency in the production of energy, even the utilization of energy deserves attention. A better utilization of energy at the public, industrial, collective, and individual levels can provide major benefits in CO_2 emission reduction.

At the public level, several authorities all around the world have put limits to the temperature within buildings during the cold and warm seasons, reducing the fossil fuel consumption and the emission of CO_2. Overall, industries consume roughly one-third of the total energy (21% of the heavy industries) with some 50% lost as waste heat. All over the world, industries are implementing saving measures with heat recovery from flue gases or fluids and reuse either internally or externally, depending on the temperature, a fundamental factor in deciding the use of heat. Usually, waste heat is divided into three segments, according to the temperature, namely, below 200 °C (*ca.* 35% of the total amount of waste heat, at the EU level), which is often referred to as low-temperature waste heat; in the range 200–500 °C (*ca.* 25% of the total); and above 500 °C (*ca.* 40%, mostly in the range 500–1000 °C). The temperature of waste heat depends on the specific industrial sector within which heat is generated [6]. For example, the higher temperature range (500–1000 °C) is typical of iron, steel, and cement industries. The 200–500 °C range is proper of the pulp and paper, chemical/petrochemical, and iron-steel industry, and the low range (<200 °C) is common to most light industrial sectors, including food. Concerning the use, high-temperature heat (>200 °C) can be used for producing high-temperature and high-pressure steam for driving turbines and producing electric energy, or for pre-heating fluids used in industrial processes. Conversely, low-temperature heat (150–200 °C) can be used for heating water used in turn for heating greenhouses (for growing vegetables and flowers) or even civil and industrial buildings and houses.

At the collective and individual level, measures should be taken to avoid misuse of energy. In the day-to-day life, each of us can adopt attitudes that would be very useful to all. Switching lights off in an empty room, avoiding superheating and supercooling of private buildings, avoiding seasonal heating (cooling) of interiors with open windows and doors, using domestic washing machines with full load, and avoiding use of electricity at any time is possible.

The transport sector is much improving the use of energy with a number of measures for upgrading the performance of cars, trucks, and buses. Lighter materials, aerodynamics, low rolling resistance of tyres, and several other smart solutions may increase the fuel mileage that is nowadays more than doubled with respect to 30 years ago. Even engines have been much improved with continuous transmission direct injection, stop and go, turbo-injection, valve timing, and cylinder deactivation that reduces the fuel consumption and, thus, emissions. Driving behaviors also impact the fuel economy. Using collective transport means instead of using personal cars and whenever possible sharing cars reduce the amount of fuel burned. Removing unneeded accessories (i.e., roof racks) and cargo, actioning smooth acceleration and deceleration, and use of high gears when at a steady speed are some examples which reduce fuel consumption. The driving behavior that perhaps gives the largest potential contribution to fuel efficiency, across nearly all vehicle types, is idle reduction, especially in cities. Idling for long periods of time wastes fuel and poses a health risk to people in and around the vehicle. For example, a typical idling bus or even school bus diesel engine burns about 2–3 L of fuel per hour. Vehicle diesel engines are generally tuned for optimum operation at traveling speeds, so they combust fuel less efficiently when idling. Idling results in more pollutants per liter, and these pollutants are concentrated in one place because the vehicle is stationary. These are typically places where vulnerable people, such as school children, are gathered. Diesel exhaust contains particulate matter which lodges in lung tissue when inhaled and is believed to cause or exacerbate numerous health problems, including cancer, asthma, reduced lung function, and premature death.

4.5 Quasi-zero-Carbon Emission Sources of Energy: Use of Biomass

In the last years, there has been a world campaign promoting so-called "*zero CO₂ emission*" fuels. They are derived from a variety of biomass, said to be formed from atmospheric CO_2 (see Fig. 2.1): the general view is that the CO_2 formed in the combustion will be fixed back in the newly formed biomass. Such *biofuels* can cover the full range of fuels derived from fossil-C and, in principle, can substitute: solid fuels (wood instead of coal), liquid fuels (ethanol and long-chain hydrocarbons instead of gasoline and long-chain fatty esters instead of diesel), and gaseous fuels (biomethane replacing natural gas). It is worth considering that wood has a quite lower energy density with respect to coal, while liquid biofuels and biomethane match quite well with the analogous fossil fuels. One of the conditions with biofuels is that the use of biomass as source of fuels must not conflict with its use as food or feed. This point has been debated for long time and, nowadays, second- and third-generation biomasses are used that do not conflict with food. Nevertheless, the use of arable land for growing biomass for food/feed or energy is still a point of argument. Marginal lands are today evaluated for growing non-food biomass together with the use of non-drinkable water. Such marginal lands can be either polluted soils not suited for growing crops because biomass can uptake pollutants from soil and transfer them to the food chain, or poor soils (low-carbon/nitrogen, arid) which would not guarantee an intensive production of edible biomass and increase the cost. The efficient use of biomass is a must and the recently developed biorefinery approach is the most valuable strategy [7].

Biofuels, such as bioethanol, long-chain hydrocarbons, biodiesel (esters of fatty acids), and biomethane, are today used at different rates in various parts of our planet. They are derived from a variety of biomasses, which have different production costs and impacts [8]. Biofuels are said "*zero-emission*" fuels, a belief partly supported by Life Cycle Assessment (LCA), a useful environmental appraisal methodology, which, nevertheless, is essentially static and needs to be adapted to the dynamic nature of biosystems. LCA studies often suffer the definition of system boundaries, the quality of available data, the completeness of data, the normalization of data, the attempt to use LCA in an absolute, instead of comparative, mode. All biofuels *are "not* really *zero-emission"* because biofuels do not offset combustion CO_2 emissions.

There are at least three key points that must be taken into due consideration for a correct assessment of biofuels. A first key point is that biomass uses carbon existing in land for growing and at the end of the day the change of soil from a baseline should be taken into due consideration in assessing the carbon balance using bio versus fossil fuels. In a wild environment, the carbon content of soil is kept almost constant because dry-fallen-down biomass re-enters soil. In a managed land, harvesting biomass makes the soil depauperated of carbon. Therefore, the amount of carbon in soil decreases and this amount is usually not considered in assessment studies. In a sense, biofuels, as fossil fuels, even if at a reduced rate, are moving carbon from soil to the atmosphere: fossil-C is taken from deep deposits and biomass is made from surface carbon. A different methodology should be used for a correct assessment of biofuels and the real contribution they can give to reducing the amount of carbon transferred from soil to the atmosphere and the overall increase of GHGs. A methodology that takes into account the amount of carbon in soil, the so-called Annual Basis Carbon (ABC) [9], has been proposed which, integrated with LCA, may give a more correct assessment of the potential of biofuels in mitigating CO_2 emissions. This potential is also limited by the fact that biomass for producing biofuels demands water: the water footprint of biomass is a second key factor that is quite important [10]. Water footprint is divided into three classes: blue water (freshwater), green water that evaporates from soil, and gray water or processed wastewater [11]. In addition, we must consider that in order to keep high the productivity of land, N-based fertilizers are used that emit N_2O. In the long term, the contribution of N_2O to GHE will increase and become more important than CO_2 due to its higher GWP. A positive news is that during last 20 years the overall amount of energy for producing biomass has significantly decreased with the implementation of more effective farming techniques (reduction of the use of fertilizers and pesticides, of the energy used in seeding, growing, and harvesting) and biomass processing [12].

As a third key point, we have to consider, as already discussed, that the combustion rate of biofuels is much higher than the fixation rate of CO_2 into biomass and this will make that yearly there will not be a closed cycle between the emitted CO_2 and the fixed CO_2, unless unlimited surface areas are considered for the production of biomass. The claimed "*zero emission*" is a wish, not a reality, at least for now. Moreover, the economics of producing bioethanol from waste biomass are negative even if oil is quoted at 100^+ US\$/barrel: fossil fuels are very cheap [13].

Still a lot of work has to be done for making biofuels environmentally and economically convenient with respect to fossil fuels. The IEA organization has published a "*technology roadmap*" on transport and biofuels [14] according to which second-generation biofuels will rise their share after 2020 and play a key role toward 2050 reaching 27% of the total fuels used in transport [15].

Recent studies confirm that it is possible that biofuels will reduce the GHG emission with respect to fossil-C, but this cannot be a general assumption, instead it demands very special conditions. In any case, the question of "*payback time*" remains an open question [16].

Table 4.2 shows the trend of biofuel use in the world since 2004 until 2030. There is clearly a continuous growth, according to the IEA studies [17].

Today, Brazil is the country that most uses biofuels (bioethanol): it is expected roughly double its consumption. Major increments are foreseen for OECD countries, EU, and developing countries and which biomass will expand more will depend on a number of factors and there is not a unified view as for today. Most likely, bioethanol will remain the major player among biofuels, considered that it can be produced from practically any cellulosic biomass (grown and residual).

The expanded use of waste/residual biomass will greatly improve the emission mitigation power of biofuels. In fact, such biomass fraction ends on as CO_2 upon burning or decomposition in soil.

Its conversion into biofuel will allow avoiding the additional burning of other C-based materials and most of the energy stored in it can be advantageously used.

The increased contribution of biofuels will change significantly the energy mix in future years, as shown in Table 4.3.

As Table 4.3 shows, oil consumption will be almost halved with respect to today, while gas and coal will decrease by 10–15%. Biofuels will more than double their contribution. Perennial sources (solar, wind, hydro, and geothermal) will be discussed in Chap. 5.

Such trend will impact positively the emission of CO_2 that will be reduced by some 10 Gt/y with respect to today [15].

Table 4.2 Trend of use of biofuels in the world (2015–2030). Figures are in Mt$_{oe}$

Entry	Country(ies)	2015	2030	% increase
1	**OECD**	**39.00**	**51.80**	133
2	North America	20.50	24.20	
3	*United States*	*19.80*	*22.80*	
4	*Canada*	*0.70*	*1.30*	
5	**Europe**	**18**	**26.60**	147.8
6	**Pacific**	**0.40**	**1.00**	250
7	**Transitional economies**	**0.10**	**0.30**	300
8	Russia	0.10	0.30	
9	**Developing countries**	**15.30**	**40.40**	264
10	Developing Asia	3.70	16.10	
11	*China*	*1.50*	*7.90*	
12	*India*	*0.20*	*2.40*	
13	*Indonesia*	*0.40*	*1.50*	
14	Middle East	0.10	0.50	
15	Africa	0.10	3.40	
16	*North Africa*	*0.10*	*0.60*	
17	Latin America	10.40	20.30	
18	**Brazil**	**10.40**	**20.30**	195
19	**World**	**83.2**	**140.4**	168

Table 4.3 Share of energy sources (as Mt$_{oe}$) until 2050 [18]

Energy resources	Year					
	2000	2010	2020	2030	2040	2050
Oil	3 575	3 981	4 264	3 849	3 178	2 433
Natural gas	**2 177**	**2 868**	**3 424**	**3 627**	**3 390**	**2 895**
Coal	2 343	3 469	4 322	4 755	4 514	3 797
Nuclear	**584**	**624**	**673**	**825**	**950**	**1 052**
Hydro	602	784	1 000	1 200	1 400	1 600
Wind	**7**	**78**	**319**	**592**	**868**	**1 145**
Solar	0	7	180	578	1 008	1 440
Geothermal and biomass	**45**	**83**	**145**	**195**	**245**	**295**
Biofuels	0	60	100	150	200	250
Total	**9 342**	**11 956**	**14 427**	**15 771**	**15 754**	**14 906**
World GDP	61 345	88 736	122 160	150 510	167 213	174 561
GDP per capita ($)	**10 053**	**12 888**	**15 912**	**17 957**	**18 595**	**18 371**
	6 567	7 422	8 466	9 533	10 592	11 671
Carbon dioxide emissions (Mt)	25 931	32 876	38 253	39 169	35 600	29 309

Oil consumption excludes biofuels. Mt: million metric tons; Mtoe: metric tons of oil equivalent
$/t_{oe}$ constant 2011 international dollars per metric ton of oil equivalent

DI4.1: **Innovation in Coal to Power (CtP) and Gas to Power (GtP): Decarbonization of fossil fuels prior to combustion. The IGCC technology**.

In a standard power plant, coal (or natural gas) burns to heat its boilers to about 540 °C to create high-pressure steam. The steam is piped to the turbines at pressures of more than 12.5 MPa.

The turbines are connected to the generators and spin them at 3 600 revolutions per minute (rpm) to make alternating current (AC) electricity at 20 000 V. Soft (river) water is pumped through tubes in a condenser to cool and condense the steam coming out of the turbines.

An average plant may generate about 10 billion kWh/y, or enough electricity to supply 700 000 homes. To meet this demand, 14 000 t of coal a day should be burned.

A old style C(NG)tP plant has an efficiency of 28–33% in the conversion of chemical energy into electricity.

Since 1985, the efficiency of electricity production has been increased by *ca.* 20%. Coal to Power (CtP) or Natural Gas to Power(NGtP), all together C(NG)tP, modern plants have an efficiency of 46$^+$%, with an important reduction of CO_2 emission.

The improved technology works with a steam temperature of over 600 °C and a pressure over 28 MPa. Additionally, heat exchangers recover some of the heat that is still contained in the flue gases that is used to preheat the combustion air and the water circulating in the water-steam circuit, thus increasing the overall efficiency of the plant by *ca.* 20$^+$%.

Another important improvement is represented by the innovative technology for using lignite (very abundant in some geographical areas) instead of coal. Lignite has a higher hydrogen content with respect to coke and a lower energy density (see Table 1.5). Drying lignite at above 100 °C using residual heat of the power production decreases its water content (hydrogen and oxygen) and increases the carbon content making it a better fuel. In this way, steam is produced at over 580 °C and 27.5 MPa, close to the improved coke plant, resulting in a great economic benefit for countries which are rich of poor-quality coal.

The use of heat of flue gases for superheating steam up to 600 °C and pressurize it at P > 28 MPa is a plus that improves the turbine work and increases the electricity production.

A further improvement on the coal to power technology is represented by the so-called IGCC or integrated gasification combined cycle. In such technology, coal is converted first into syngas which is improved by combining with the WGS reaction. At the end, a mixture of H_2 and CO_2 is formed accompanied by some other impurities (SOx, CO, and others) (Eq. 4.1).

$$C + 2H_2O \rightarrow CO_2 + 2H_2 \tag{4.1}$$

CO_2 is separated by using ethanolamine or membrane technologies and H_2 is burned with oxygen. CO_2 having been separated before combustion, the flue gases are made essentially by water vapor that can be eventually condensed and reused. The collected CO_2 can be purified for the end use.

Such technology produces electricity by driving a high-efficiency gas turbine, and the exhaust heat is recovered to produce steam to power traditional high-efficiency steam turbines (Coal Utilization Research Council) [19]. IGCC can use pulverized coal and natural gas as feed, even bituminous coal and petroleum coke have been cofed. Even if IGCC has an efficiency which is some 12–14 points % higher than a conventional upgraded PCC plant, its CAPEX are some 20% higher and even OPEX are higher than those of conventional plants due to the highly sophisticated technology. Six plants are operated and more are under construction. Nevertheless, the real potential of IGCC is not easy to foresee.

Table 4.4 Coal-fired IGCC demonstration plants

Plant	Location	Output (MWe)	Feedstock	Gasifier	Operation
Tampa Electric Polk Plant	Polk County, FL, USA	250	Bituminous coal and pet coke	GE	1996–Present
ConocoPhillips Wabash River Plant	**West Terre Haute, IN USA**	**262**	**Bituminous coal and pet coke**	**E-Gas®**	**1995–Present**
NUON/Demkolec Willem-Alexander	Buggenum, The Netherlands	253	Bituminous coal and biomass	Shell	1994–Present
ELCOGAS/Puertollano	**Puertollano, Spain**	**318**	**Coal and petroleum coke**	**Prenflo®**	**1998–Present**
CCP	Nakoso, Japan	220	Bituminous coal	MHI	2007–Present
Vresova	**Czech Republic**	**350**	**Coal/lignite**	**Lurgi, Siemens**	**1996–Lurgi 2008– Siemens**

More recently (2016), another plant was built in the Kemper County, USA. Not all the seven plants are working at their full potential. One can estimate that only four are continuously performing at their best level in USA and EU. However, budgetary issues are slowing down the IGCC spreading, as it requires high CAPEX, despite it has a beneficial impact on the environment as it reduces the emission of CO_2 per unit electric power. Moreover, as CO_2 can be separated before combustion occurs, it is cleaner than that produced by post-combustion recovery from conventional power plants. In fact, it does not contain NOx and is very poor of S-derivatives. If environmental costs are taken into the due consideration, IGCC may become the technology of the future.

China, India, EU, and USA are among the most interested in supporting such technology. As IGCC goes through syngas production, then it is not a standing alone technology but can be integrated into a network of technologies: hydrogen production/utilization, FT, synthetic fuels, and synthetic chemistry.

Optimistic views are not absent in the global panorama, at least for what concerns the OPEX costs [20]. Such study, old of 10 years now, has compared the cost of several energy-production technologies to the IGCC cost. Interestingly, the study affirms that OPEX for nuclear plant with advanced reactor designs would be £1 200/kW (1 £ = ca. 1.3 US$), with the cheapest eolic option at £1 070/kW for an onshore wind farm and £1 375/kW for offshore wind tower: such values are comparable to the cost of IGCC without carbon sequestration, set at £1 250/kW. The price of CO$_2$ capture and storage would be strongly dependent on the geographic area and the distance of the production from the storage site. Renewables (biomass) and solar were estimated to be more than double the IGCC cost per kilowatt installed in the same study.

If IGCC will guarantee clean electric energy, it will result to be a strong support to the use of coal, the most abundant fossil-C we have.

References

1. https://commons.wikimedia.org/wiki/File:Historical_CO2_levels_based_on_proxy_(indirect)_measurements.png
2. a) Friedman A (2013) IPCC report. https://www.climatecentral.org/news/ipcc-climate-change-report-contains-grave-carbon-budget-message-16569; b) Cointe B, Ravon PA, Guerin E (2011) 2°C: the history of a policy-science nexus. Working papers n° 19/11. Institute for Sustainable Development and International Relations, Paris. https://doi.org/10.13140/RG.2.1.1876.0564
3. https://portals.iucn.org/library/efiles/documents/EPLP-086.pdf
4. a) https://www.mpoweruk.com/energy_efficiency.htm; b) https://greatercea.org/lightbulb-efficiency-comparison-chart/
5. https://www.researchgate.net/publication/225651919_Energy_efficiency_technologies_for_road_vehicles
6. Papapetru M, Kosmakadis G, Cipollina A, La Commare U, Micale G (2018) Industrial waste heat: estimation of the technically available resource in the EU per industrial sector, temperature level and country. Appl Therm Eng 138:207–216
7. Aresta M, Dibenedetto A, Dumeignil F (2012) Biorefinery from biomass to chemicals and fuels. De Gruyter
8. Salassi ME, Brown K, Hilbun BM, Deliberto MA, Gravois KA, Mark TB, Falconer LI (2014) Farm-scale cost of producing perennial energy cane as a biofuel feedstock. BioEnergy Res 7:609–619
9. De Cicco JM (2018) Methodological issues regarding biofuels and carbon uptake. Sustainability 10:1581–1596
10. Hammond GP, Li B (2016) Environmental and resource burdens associated with world biofuel production out to 2050: footprint components from carbon emissions and land use to waste arisings and water consumption. Glob Change Biol Bioenergy 8:894–908
11. Hammond GP, Seth SM (2013) Carbon and environmental footprinting of global biofuel production. Appl Energy 112:547–559
12. https://www.eere.energy.gov
13. https://www.researchgate.net/publication/271040257_A_Review_of_Life_Cycle_Assessment_LCA_of_Bioethanol_from_Lignocellulosic_Biomass
14. IEA, Technology roadmap: biofuels for transport. OECD/IEA
15. Aresta M, Dibenedetto A, He N (2013) Report for the catalyst group

16. Elshout PMF, van der Velde M, van Zelm R, Steinmann ZJN, Huijbregts MAJ (2019) Global relative species loss due to first-generation biofuel production for the transport sector. Global Change Biol Bioenergy 11(6):763–772
17. Gupta D, Kumar Gaur S (2019) Carbon and biofuel footprinting of global production of biofuels. In: Biomass, biobased-based materials, and Bioenergy, pp 449–481
18. Patterson R (2016) World energy 2016–2050: annual report. Political Economist. https://content.csbs.utah.edu/~mli/Economies%205430-6430/World%20Energy%202016-2050.pdf
19. Suarez-Ruiz I, Crelling JC (eds) (2008) Applied coal petrology: the role of petrology in coal utilization. Academic Press
20. Jeffs E (2008) In: Generating power at high efficiency. Elsevier

The Alternative, Carbon-Free Primary Energy Sources and Relevant Technologies

5

Abstract

In this chapter, carbon-free perennial primary energy sources are described and the potential of relevant technologies to produce electric or thermal energy is highlighted. Solar, wind, hydro, and geothermal power are analyzed for their contribution to substitute fossil-C in coming decades.

5.1 Carbon-Free Primary Sources Alternative to Fossil-C and Biomass

The use of biofuels, and man-made fuels (see Chaps. 9–11), may contribute to reduce the CO_2 atmospheric burden by cycling CO_2 and avoiding the extraction of fossil-C. As discussed in Chap. 4, the biofuels' contribution to the mitigation of CO_2 is not straight to evaluate and cannot be measured on the basis of substituted fossil fuels. As a matter of fact, only upon adopting innovative culture practices, advanced harvesting techniques, and processing technologies, one can expect a positive contribution from biofuels. In a Business as Usual (BAS) frame, the risk is that biofuels may increase the emission of CO_2 with respect to fossil-C. Moreover, a limitation is set by the kinetics of biomass growing that can be improved only to a marginal extent. Alternative to the use of biomass is the use of *C-free perennial energies*. Such energy sources are connatural to our planet and always available. Man, since ever, has made attempts to use them in the best of ways. We refer to

- Solar energy,
- Wind energy,
- Hydroenergy (internal waters and oceans), and
- Geothermal energy.

© Springer Nature Switzerland AG 2021
M. Aresta and A. Dibenedetto, *The Carbon Dioxide Revolution*,
https://doi.org/10.1007/978-3-030-59061-1_5

Such primary energy sources are C-free and can, thus, offset CO_2 emissions. There are issues linked to their use, derived from their cyclical availability (day–night cycles of solar energy), discontinuity (wind), and geographic availability (hydro and geothermal energy). Nevertheless, man has since ever used such sources with a variable intensity dictated by the easier availability of other energy sources. Fossil-C has the great advantage of having the highest energy density and being easily transported everywhere in the world. Such plus is not easy to even up. As a matter of fact, differently from the various forms of fossil-C and biomass that can be shipped to any part of the globe from any part and will have everywhere the same energetic density and intensity, the above perennial energies can have only a local utilization. It is one of their derivatives that can be transported and transferred. And among the derivatives (heat, electrical, and mechanical energies), it is the electric energy that most suited for distribution, with all its own limitations. If oil, gas, and coal can be shipped from deep south to the upper north without problems, electricity cannot yet be transported from Australia to Iceland. Maybe in future *highways to electrons* will be open that are not too dissipative; today, the only way is the use of cables or the more expensive and less intensive batteries. The discontinuous availability of the primary sources, their geographical availability, and the variable local intensity are issues that must be engineered for an optimal and wide utilization of such sources. We shall now consider singularly the perennial sources listed above and give the frame into which they can better contribute to the reduction of extraction and use of fossil-C.

5.2 The Use of Perennial Energy Sources

5.2.1 Solar Energy

Solar energy is by far the most abundant source of energy we have. It is estimated that 1 000 W/m^2 reach the Earth surface [1] each day with the total energy reported in Entry 1, Table 5.1 that compares the various forms of energy available on Earth.

Table 5.1 Comparison of various forms of perennial energy on Earth with biomass	Entry	Energy source max.	Power (*TW*)
	1	**Total surface solar**	**85 000**
	2	Desert solar	7 650
	3	**Ocean thermal**	**100**
	4	Wind	72
	5	**Geothermal**	**44**
	6	River hydroelectric	7
	7	**Biomass**	**7**
	8	Open ocean wave	7
	9	**Tidal wave**	**4**
	10	Coastal wave	3
	11	**Total Energy used in 2019**	**15**

Solar energy can be used for generating heat at various temperatures or electricity. The former use is the simplest and known since long time, and electricity production is a much more recent achievement.

5.2.1.1   Heat Generation

Solar energy can be used for heating a fluid (either air or a liquid) that is then used for a variety of applications. *Flat collectors* are very popular, in which water, eventually added with anti-freeze agent (ethene or propene glycol, non-toxic, non-corrosive) to prevent icing during cold days, is warmed up. In general, such solar heating systems can be either active (once water reaches the wished temperature is circulated to a collector with the use of a pump) or passive (without pump). The active systems can be either direct (water is heated) or indirect (an anti-freeze agent is warmed-up which in turn transfers heat to water). The latter is more suited for areas where it can freeze. Such heaters work all year long. An alternative to flat collectors are the so-called *integral collectors* made of a cylinder where water is preheated and then circulated to external tubes for continuing heating. Such systems are suited for non-icing areas. A third device used is the *evacuated-tube solar collectors* made of transparent tubes whose internal surface is covered with materials which prevent loss of heat by radiation. Heat is absorbed by metal elements. Such heaters are suitable for houses and some industrial buildings. The heat generation unit is coupled to a backup unit for providing heat during dark and in case of over demand. In addition to such devices which operate at mild temperature, sun energy can be even collected and concentrated to reach high temperature (>1000 °C).

Concentrators of solar energy (Fig. 5.1) can be of various types [2], namely, parabolic, hyperboloid, Fresnel, CPC, dielectric totally internally reflecting, flat

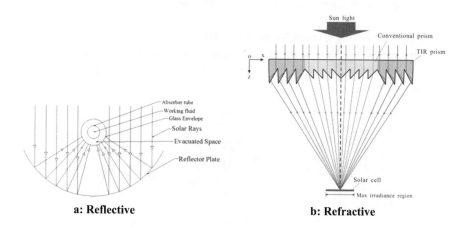

a: Reflective   **b: Refractive**

Fig. 5.1 Concentrators of solar energy: solar rays are focused to the focal point of the parabola. **a** Reprinted from Ref. [2a] Copyright © 2016 with permission from Elsevier; **b** Reprinted from Ref. [2b] Copyright © 2014, with permission from Elsevier

high concentration, and quantum dot. Most of such systems demand a very good sun-tracking system.

We shall further consider two types of devices: the parabolic or reflective concentrators (Fig. 5.1a) and Fresnel or refractive concentrators (Fig. 5.1b). Fresnel lenses have various advantages but are more difficult to construct. The receiver may have a variety of structures and shapes: cylindrical, flat plate, even a cavity, etc.

Solar concentrators have been built at various sites in USA, Spain, Israel, Germany, and Australia using the modes described above. The efficiency is highly variable from 12 to 81%. They are characterized by a variety of geometries: (A) conical concentrator, (B) CPC concentrator, (C) sphere concentrator, (D) cylinder array concentrator, (E) array Fresnel lens concentrator, (F) heliostat tower concentrator, (G) parabolic concentrator, (H) array parabolic concentrator, (I) reflective Fresnel lens concentrator, (J) small disk reflective concentrator, (K) convex lens concentrator, (L) transmittance Fresnel lens concentrator, (M) trough conical concentrator, (N) mirror dual-focus concentrator, and (O) multi-curved compound concentrator.

The concentrated solar energy can be used for a variety of purposes. High temperature can be used to run "energy demanding or endoergonic" chemical reactions or for producing steam for running a turbine that will generate electricity, or for producing high-temperature fluids that can work as solar energy storage, etc. An interesting application is the production of diesel from air (vide infra). Among thermal energy storage (TES) devices, we mention molten salts [3] that exist in the likely range 150–600 °C, with some high levels of 250–1000 °C. They can be used for storing solar energy collected during daylight and produce high-temperature and high-pressure steam for generating electric energy; in this way, a continuous electron flow can be generated day and night, offsetting the intermittency of solar light that would be a serious problem in a frame in which solar energy would be used for feeding electricity to the network (vide infra).

5.2.1.2 Electricity Generation: Photovoltaic Devices

A second way to use solar energy is the direct conversion into electricity by using a photovoltaic device (Fig. 5.2) [4]. A photovoltaic cell is made up of "*semiconducting materials.*"

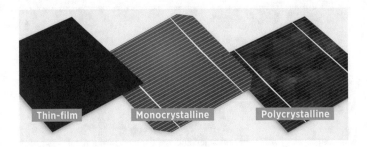

Fig. 5.2 Examples of solar cells [4]

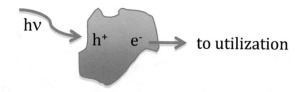

Fig. 5.3 Conversion of solar energy into electricity using semiconductors

A semiconductor is a material that when irradiated by light generates a *"charge separation,"* e.g., a *"hole–electron, h-e⁻"* couple, also known as an *"exciton"* (Fig. 5.3). Electrons flow in the utilization circuit.

Several materials are used for converting solar light into electricity. *Crystalline Silicon* (Si) is by far the most used for building PV cells for its low cost and durability (25 + years). *Thin films* are an alternative. In this case, photoactive materials are deposited as thin film on a support such as a glass or plastic surface. Most used photomaterial for these films is Cadmium Telluride (CdTe). Even Copper Indium Gallium Diselenide ($CuInGaSe_2$) is used. Such thin films are less resistant than Si and need protection when used outdoor. A third class of photovoltaic materials are Organic Photovoltaic (OPV) [5]. Such materials are cheap and easy to manufacture, but have a shorter life than Si-based PV. Nevertheless, they are able to use specific wavelengths and can be built more easily in bulk amounts than crystalline Si. Finally, Concentration Photovoltaics (CPV) have been developed in which solar light is concentrated on the cell and this increases the efficiency of light conversion. Such technology has the drawback of being more expensive.

During last decade, the cost of production of PV cells has decreased (halved in the last 10 years) while efficiency in solar light conversion has increased (up to 20–24%, with a forecast of reaching 40% in near future). Lowering of price and increasing of conversion efficiency guarantee a future to PV that will rise from actual expected 650 GW (2019) to 3 500–4 500 GW by 2040 (Fig. 5.4).

A single PV cell produces an *"Open-Circuit Voltage"* (V_{OC}) of about 0.5–0.6 V at 25 °C (typically around 0.58 V), no matter how large it is. This cell voltage remains fairly constant just as long as there is sufficient irradiance light from dull to bright sunlight (range 1–10 W/m²). Open-circuit voltage means that the PV cell is not connected to any external utilizer and is therefore not producing any current flow. A panel of 400 cells will generate a $V = 400 \times 0.58 = 232$ V, that is, the voltage used in civil buildings, public and private. Usually, modules of 36 cells that deliver 21 V peak and 100 W are assembled to reach the desired voltage and power. A first constrain to maintain the maximum power output and efficiency of a **photovoltaic panel** is that the PV panel must constantly face the sun. This can be easily achieved by using a simple technique called *Solar Panel Orientation* to automatically track the movement of sun across the sky during daylight, or by manually setting the angle of the PV panel toward the sun and then adjusting it.

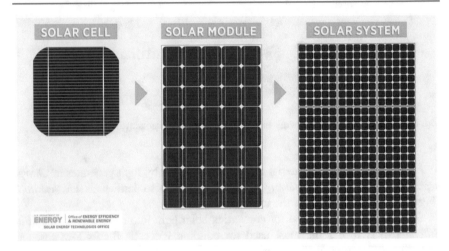

Fig. 5.4 From a *solar cell* to a *solar system* [6]

A second condition is that the surface of the panel is kept clean: this obliges to place the panel in such a position that cleaning is easily done (Fig. 5.4).

Solar energy conversion to electricity can be done in small personal units for serving a family on in open fields connected to the electricity transfer network. Even in the first case, the modules are connected to the network where they can send electricity not used during daylight and from which the private utilizer can take electricity during peak time or night.

Use of PV is expanding and the total power installed is growing at an interesting rate and new technologies (ultralight and flexible panels) are producing modules that can be adapted to a variety of utilizations.

5.2.2 Wind Energy

Wind, a different form of solar energy, has been used by humans since very long time in a variety of activities: from cruising seas and oceans to cool down buildings and houses to application in productive activities. Until the end of 1800s, windmills were largely used for generating power for several activities: sailing, crushing goods, grinding grains (making floor), pumping water, and making electric energy. Wind is discontinuous and site specific and was left over after fossil fuels were used for generating electric energy.

Nowadays, wind turbines [7] are back in use both on land and offshore for producing electric energy. Even wind towers, [8] typical of the Persian architecture, are coming back in use as passive non-electricity-demanding devices for cooling civil buildings. Concerning electricity, the wind total installed capacity was 591.5 GW in 2018 and expected to reach 655.9 GW in 2019.

Wind and solar power are very close and grow at a similar rate: in 2019 with respect to 2018, wind-installed power has grown by 27.4% and solar by 34.3%.

The way wind turbines work is quite simple and recalls the generation of electric energy in a power station; in both cases, the protagonist is a flowing fluid that operates a turbine connected to a generator. In a power station, it is the high-temperature and high-pressure steam generated by burning fossil-C the driving force, and in wind turbines it is the kinetic energy of air (wind) that forces the blades of the tower mill to run a turbine which spins a generator of electricity.

There are two kinds of wind turbines: a horizontal axis (Fig. 5.5, left) and a vertical axis (Fig. 5.5, right) [7].

Vertical axis turbines have the advantage to be omnidirectional, which means that they work whatever is the direction of blowing of wind and do not need to point to the wind as in the case of horizontal-axis turbines.

The power of wind turbines varies from 100 kW (personal use) to 2–3 MW. Typically, a wind turbine has three blades each of length 52 m which swept a surface of 8 452 m^2 and a theoretical power generation of 3.6 MW. Wind turbine only seldom work at their maximum capacity: several factors influence their power generation, such as wind speed, direction, and continuity.

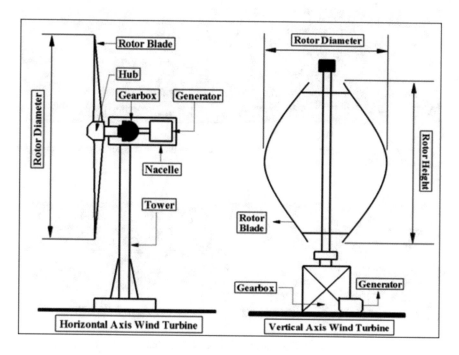

Fig. 5.5 Operating wind turbines. Left: horizontal axis; Right: vertical axis. Reprinted from Ref. [7], Copyright (2015), with permission from Elsevier

5.2.3 Hydro Energy (Both Internal Waters and Oceans)

Hydroelectric energy has a long story. The kinetic energy of running water, similarly to that of wind, has been captured and used for many different purposes: milling was very popular since ca. 600 bC. Figure 5.6 shows an old (ca. thirteenth century) watermill [9]: running water moves a wheel with blades which is connected to an axis (left) that transfers the movement to other mechanical parts such as a milling machine.

On the right, the movement is transferred by using pulleys, and in other cases gears are used. The more modern use of water power is the production of electric energy, which follows a scheme similar to the windmills. Electric energy is produced on a large scale by either directly using running water (river) or by collecting water in a basin and using "*dams*" (Fig. 5.7) that can be opened or closed, to regulate the flow of water to the generator [10].

With respect to the direct use of running water, the basin–dam system allows a more controlled, constant, and continuous delivery of kinetic energy to the generator. What has to be emphasized is that the production of electric energy from water kinetic energy is very efficient: over 90% of the water kinetic energy is converted into electric energy! This figure has to be compared with the value of

Fig. 5.6 An old (*ca*. thirteenth century) water mill (left) and the transmission of movement (right) using pulleys [9b]

Fig. 5.7 The operative scheme of generation of electric energy (left) and the view of a dam (right) [10]

50% in best power plants fed with fossil-C. The reason is that with the use of hydropower, it is saved the loss of proper energy of the combustion of fuels and generation of high-temperature and high-pressure steam. Moreover, hydroelectricity is very clean: no emission of toxic or polluting species occurs. The cost of production (excluding CAPEX) is around 0.01 US\$/kWh, compared to 0.032–0.035 US\$/kWh of fossil-C-based power plants. The difference is due to the fact that "*fuel*" (water) is free [11]. The world's largest hydroelectric plant in terms of installed capacity is Three Gorges (Sanxia) on China's Yangtze River, which is 2.3 km wide and 185 m high. Conversely, the Itaipu plant situated on the Paraná River between Brazil and Paraguay generates the most electricity annually. As said above, hydroelectricity is a clean and relatively cheap form of energy. OPEX costs are very low, "*fuel*" is free (river water). The power produced can be regulated with water falling on the turbines. What is negative is the environmental impact of building the basin and the dam. However, big dam projects can disrupt river ecosystems and surrounding communities, harming wildlife and forcing out residents. The Three Gorges Dam, for example, displaced an estimated 1.2 million people and flooded hundreds of villages. Hydroelectric power is not ubiquitous and depends on the existence of rivers and on the orography of the territory.

A different form of hydro energy that can be used for generating electric energy is the "*tidal energy*" [12] that is not yet widely used, but has potential for future electricity generation.

The advantage of tidal energy is that it is clean, is more predictable than the wind and the sun, and is continuous. In fact, the gravitational forces of celestial bodies will not stop nor will have unforeseen delays. However, it is far easier for engineers to design efficient systems, than, say, predicting when the wind will blow, its direction and speed, or when the sun will shine and the intensity. As for today, the largest tidal project in the world is the 2011 Sihwa Lake Tidal Power Station in South Korea, with an installed capacity of 254 MW. The project took advantage of a 12.5 km wall built in 1994 to protect the coast against flooding and to support agricultural irrigation. Just for comparison, the Roscoe wind farm in Texas, US, the first single wind park in the world, second only to the multiple wind park Gansu in China that generates 5.1 GW, takes up 400 km^2 of farmland and generates 781.5 MW with 627 wind turbines.

Tidal turbines can generate electricity at speeds as low as 1 m/s; in contrast, most wind turbines begin generating electricity at 3–4 m/s. Even if today tidal power converters may appear costly for their output, technological innovation will only drive it cheaper and more sustainable. A rough estimate sets at ca. 10–20% of global electricity demand the contribution of wave power. Tidal power plants can last much longer (over 100 years) than wind or solar farms, which come with a warranty of 20–30 years (some have reached 40 years) and are reported to degenerate at a yearly rate of ca. 0.5% efficiency. The La Rance tidal power plant in France has been operational since 1966 (53 years now, with a old technology) and continues to generate significant amounts of electricity each year. Tidal barrages are long concrete structures usually built across river estuaries (Fig. 5.8). The barrages have tunnels along them containing turbines, which are turned when water on one

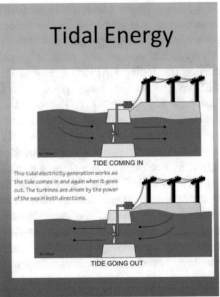

Fig. 5.8 A typical tide power plant (left) and how it works (right): water flowing through the blades causes transfer of kinetic energy to the turbine and the generator [12]

side flows through the barrage to the other side. In oceans, or large lakes, tides can be an abundant, constant, clean, and, may be in future, cheap source of energy that could be available to coastal countries which are not in reach of fossil-C, sun, and wind.

The only drawback is our scarce knowledge on the impact that tidal power plants may have on marine life. An accurate study by the European Commission says that noise, vibrations, and electromagnetic radiations may influence the marine life, even if not on a permanent mode. For example, eels were observed to be diverted from their usual migration route, but they were finally able to resume it. Electromagnetic frequency seems to attract benthic species such as sharks, rays, and skates, but not to cause detrimental damages. Deep studies are necessary in order to understand to which extent humans can take profit of this opportunity without causing damage to other species.

5.2.4 Geothermal Energy

Geothermal energy is the thermal energy of rocks and fluids on the subsurface of the Earth crust [13]. As we go deeper, the temperature increases at an average rate of 20 °C per km. Nevertheless, in some parts of the globe, such rate is much higher

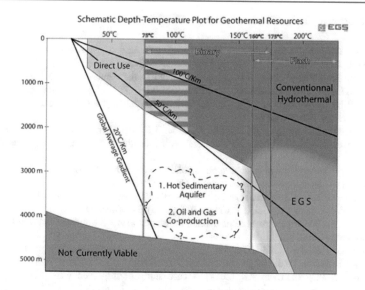

Fig. 5.9 Temperature of rocks and fluids with increasing depth [14]

and it is possible to reach 50 °C in 1 000 m. The temperature of rocks ranges over a wide interval and can reach 200 °C.

The heat at a temperature up to 80 °C can be directly used with a variety of technologies for heating (heat pumps). Heat in the interval 120–200 °C can be used for generating steam (compared with the recovery of heat from industrial plants) and, thus, electricity (Fig. 5.9).

A geothermal power plant takes advantage of hot fluids (hot water or dry steam) present at a deepness of 1.5–3 km. Dry steam springing from soil was used already in 1904 in Tuscany, Italy for generating electric power and is now popular in the entire world. A second approach is the *"flash steam."* In such technology, hot water taken from the ground is converted into steam and used to drive a turbine. Also, this technology is largely used all over the world. A third technology is the *"binary cycle"* in which hot water from the ground is used to vaporize a second liquid and such high-temperature vapor is used to run the turbine. Most recent technologies (*Enhanced or Engineered Geothermal Systems (EGS)*) use an integrated approach to use the geothermal energy. A fracture is generated in the subsurface where water is pumped and converted into steam which returns up and is used for generating electricity. Such approach is valid for areas where either the availability of water is not in abundance or rocks are not that porous. The artificial injection of water may increase the economics of a given area.

Geothermal electric power, with hydroelectric energy, solar, and wind, is a C-free energy source and has an excellent perspective of expansion. The total amount of geothermal electricity produced is 12.8 GW as for 2020.

References

1. (a) Abbott D (2009) Keeping the energy debate clean: how do we supply the world's energy needs?" Proc IEEE 98(1):42–66; (b) https://windows2universe.org/
2. (a) Jebasingh VK, Herbert GMJ (2016) A review of solar parabolic trough collector. Renew Sustain Energy Rev 54:1085–1091; (b) Zhuang Z, Yu F (2014) Optimization design of hybrid Fresnel-based concentrator for generating uniformity irradiance with the broad solar spectrum. Optic Lasers Technol 60:27–33
3. Breidenbach N, Martin C, Jockenhöfer H, Bauer T (2016) Thermal energy storage in molten salts: overview of novel concepts and the DLR test facility TESIS. Energy Procedia 99:120–129
4. https://www.energy.gov/eere/solar/articles/solar-photovoltaic-cell-basics
5. https://www.energy.gov/eere/solar/organic-photovoltaics-research
6. https://www.energy.gov/eere/solar/articles/pv-cells-101-primer-solar-photovoltaic-cell
7. Huang G-Y, Shiah YC, Bai C-J, Chong WT (2015) Experimental study of the protuberance effect on the blade performance of a small horizontal axis wind turbine. J Wind Eng Indus Aerodyn 147:202–211
8. Hejazi B, Hejazi M (2014) Persian Wind Towers: architecture, cooling performance and seismic behavior. Int J Design Nat Ecodyn 9(1):56–70
9. (a) Behari PC, Bhardwaj AK (2014) A case study of improved watermill using power electronics devices for offgrid power generation research. J Appl Sci Eng Technol 2:417–423; (b) https://en.wikipedia.org/wiki/Watermill
10. Wikimedia Commons (2015) Hydroelectric Plant [Online]. Available: https://upload.wikimedia.org/wikipedia/commons/thumb/5/57/Hydroelectric_dam.svg/2000px-Hydroelectric_dam.svg.png
11. https://www.eia.gov/electricity/annual/html/epa_08_04.html
12. https://www.power-technology.com/features/tidal-energy-advantages-and-disadvantages/
13. Goldstein B (2012) Geothermal energy, nature, use, and expectations. In: Meyers RA (eds) Encyclopedia of sustainability science and technology. Springer, New York, NY
14. Accessed from Environmental and Geothermal Services (EGS) website http://envgeo.com/publications/ on 27 April 2020

Reduction of Carbon Dioxide Emission into the Atmosphere: The Capture and Storage (CCS) Option

6

Abstract

The Capture and Storage of Carbon Dioxide (CCS) is illustrated in this chapter. CO_2 can be captured from point sources (power plants and industries) or directly from the atmosphere (Direct Air Capture (DAC)) or even from mobile sources. Technologies for the capture are illustrated here and some examples of disposal in natural fields are described, together with fixation into long-lasting materials, such as inorganic carbonates.

6.1 The Capture of CO_2

Besides the efficiency strategies in using fossil-C and the use of biomass that reduce the formation of CO_2, and the alternative C-free energy sources that avoid the formation of CO_2, scientists and technologists are working at the development of *"end-of-pipe"* technologies, which means at technologies which prevent the accumulation of CO_2 in the atmosphere. These are the technologies that *"capture CO_2,"* either before it reaches the atmosphere or even directly from the atmosphere. Once CO_2 is captured, it can be either disposed or used. These alternative aspects will be discussed in this and following chapters: they represent two faces of a medal and are relevant to two different economic views. *Disposal* of CO_2 is functional to a *linear economy*, while the *utilization* of CO_2 is at the core of *C-cyclic economy*. These are two different philosophies of life and use of resources.

Disposal demands energy for transportation and housing of CO_2, causing a penalization on the produced electric energy and, thus, an extra extraction of fossil-C in order to not reduce the available energy: the effect is shortening the availability of fossil-C. Conversion of CO_2, instead, recycles carbon, using it for making chemicals, materials, and fuels; finally, fossil reserves are less used and their availability for future generations is guaranteed. Moreover, CCS is a net cost,

© Springer Nature Switzerland AG 2021
M. Aresta and A. Dibenedetto, *The Carbon Dioxide Revolution*,
https://doi.org/10.1007/978-3-030-59061-1_6

while CCU produces goods that have an added value. CCS avoids atmospheric CO_2 by putting it somewhere underground or in deep waters, and CCU converts it back into the products from which it is generated, mimicking the natural continuous C-Cycle. *Building a man-made C-cycle that recycles carbon through CO_2 conversion is for chemistry and biotechnology a challenge that matches in complexity the fusion of nuclei in physics.* Scientists know how to divide the nucleus and how to use organized matter (biomass, coal, oil) for producing energy; making energy by nuclear fusion or catching energy in chemical bonds, is much more difficult. A bottleneck in the conversion of CO_2 into chemicals or materials is that such process is endoergic in most cases (vide infra) and may even require hydrogen. For such conversion of CO_2, energy cannot be provided by burning fossil-C nor biomass: both would emit larger amounts of CO_2 than converted [1a–d]. Therefore, recycling carbon by converting CO_2 into energy-rich species is not feasible within an energy system based on fossil-C. As we have pointed out long time ago, the conversion of CO_2 into energy products is possible only if the required energy is produced from perennial primary energy sources [2]. Chapter 5 has demonstrated that time is there: now we can! We shall discuss the capture and disposal of CO_2 in this chapter, and its use in Chaps. 8–11.

6.2 CO_2 Capture from Point Sources: Power Stations and Industrial Processes

Capture of CO_2 using a basic solution or even water is practiced since long time. In 1863, Ernst Solvay established the Solvay Company for the production of Na_2CO_3 and $NaHCO_3$. The process is based on a series of reactions starting with natural $CaCO_3$ that is thermally decomposed to afford CaO and CO_2 (Eq. 6.1). Then NH_3 and CO_2 (Eq. 6.2) react in brine (Eq. 6.3) allowing $NaHCO_3$ to set down while ammonium chloride remains in water.

$$CaCO_{3(s)} \rightarrow CaO_{(s)} + CO_{2(g)} \tag{6.1}$$

$$NH_{3(g)} + H_2O + CO_{2(g)} \rightarrow NH_4^+{}_{(aq)} + HCO_3^-{}_{(aq)} \tag{6.2}$$

$$NH_4^+ + HCO_3^- + NaCl_{(aq)} \rightarrow NaHCO_{3(s)} + NH_4^+{}_{(aq)}Cl^-_{(aq)} \tag{6.3}$$

$NaHCO_3$ is thus separated and thermally converted into Na_2CO_3 (Eq. 6.4).

$$2NaHCO_{3(s)} \rightarrow Na_2CO_{3(s)} + H_2O_{(v)} + CO_{2(g)} \tag{6.4}$$

Ammonia is recovered from ammonium chloride by treatment with $Ca(OH)_2$ (Eq. 6.6), produced by reaction of CaO (see Eq. 6.1) with water (Eq. 6.5), and recycled.

$$CaO(s) + H_2O_{(l)} \rightarrow Ca(OH)_{2(aq)} \tag{6.5}$$

$$2NH_4Cl_{(aq)} + Ca(OH)_{2(aq)} \rightarrow 2NH_{3(g)} + CaCl_{2(aq)} + 2H_2O \tag{6.6}$$

The Solvay process since over 150 years is a beautiful example of recycling of chemicals.

The capture of CO_2 has been attempted in more recent times from power plant flue gases with the intention of avoiding its entrance into the atmosphere. Basic solutions are effective means (Eqs. 6.7–6.9a, 6.9b).

$$Na(K)OH_{(aq)} + CO_{2(g)} \rightarrow Na(K)HCO_{3(s)} \tag{6.7}$$

$$Ca(Mg)(OH)_{2(aq)} + CO_{2(g)} \rightarrow Ca(Mg)CO_{3(s)} + H_2O \tag{6.8}$$

$$H_2NCH_2CH_2OH + H_2O + CO_2 \rightarrow HOCH_2CH_2NH_3^+ + HCO_3^- \tag{6.9a}$$

$$2H_2NCH_2CH_2OH + CO_2 \rightarrow HOCH_2CH_2NH_3^+ + HOCH_2CH_2NHCO_2^- \tag{6.9b}$$

$$2Na(HCO_3)_2 \rightarrow Na_2CO_3 + CO_2 + H_2O \tag{6.10a}$$

$$HOCH_2CH_2NH_3^+ + HOCH_2CH_2NHCO_2^- \rightarrow 2HOCH_2CH_2NH_2 + CO_2 \tag{6.10b}$$

Once captured in the form of hydrogen carbonate (Eqs. 6.7, 6.9a), carbonate (Eq. 6.8), or carbamate (Eq. 6.9b), CO_2 can be released by thermal processes (Eqs. 6.10a, 6.10b) or even by pressure release. Temperature or pressure swing is the technology for CO_2 recovery from flue gases. Issue with such technology is the energy necessary to recover CO_2, as it represents both an economic and environmental burden. For example, reaction (6.1) demands a temperature of 898 °C for a rapid decomposition of limestone may occur (equilibrium $P_{CO2} = 0.101$ MPa) and is endothermic with a molar enthalpy of 178 kJ/mol (or 4 045 MJ/kg_{CO2}). As a matter of fact, the decomposition of $CaCO_3$ starts above 500 °C and is completed at 898 °C when the equilibrium pressure of gaseous CO_2 equals the atmospheric pressure (0.101 MPa). Such decomposition allows the separate recovery of both the released CO_2, which has a good purity, and CaO that finds several applications among which there is even the capture of CO_2. If the capture is carried out in the solid phase, not only the kinetics will be slower, but even the process may not reach completeness as the carbonate formed on the surface of particles upon reaction of CaO with CO_2 will slow down the process or even prevent the carbonation of the inner layers. Solid reacting phases are not the most suited for an efficient and fast capture and release. Using a slurry in water may accelerate the uptake, but then will rise problems in the release step. In water, most likely the hydrogen carbonate is formed [$Ca(HCO_3)_2$] which must be dried and dewatered to afford $CaCO_3$. Using NaOH or Na_2CO_3 as sorbent in water may improve the kinetics and thermodynamics of the capture. In fact, according to reaction (6.7), NaOH in water uptakes CO_2 and affords the hydrogen carbonate which is only moderately soluble in water ($S = 11.13$ g/100 g_{H2O}). Once the solution is saturated, the hydrogen carbonate starts to set down. Similarly, Na_2CO_3 (more soluble in water than the hydrogen carbonate, $S = 28.14$ g/100g_{H2O}) can uptake CO_2 in water affording a slurry of sodium hydrogen carbonate (reverse of reaction 6.10a, 6.10b). The slurry is thermally decomposed to release CO_2 and regenerate the basic sodium carbonate solution that can be reused. For such process, the energetics is less negative than with the calcium system: 3.22 MJ/kg$_{CO2}$ captured are necessary. Alternative sorbents have been used such as ethanolamine which affords mixed ammonium carbonates (Eq. 6.9a) or carbamates (Eq. 6.9b). The overall goal in the capture process is to reduce costs (economic and energetic) and time, while keeping an eye at the security and health. The MEA process is very popular even if it has an energetics worst than the sodium carbonate system. In fact, 3.8 MJ/kg$_{CO2}$ captured are necessary, which is *ca.* 20% higher than the enthalpy of the sodium carbonate process [2]. The MEA process is at the demo-commercial scale (Fig. 6.1) [2b]. The typical operative conditions are absorption at 20–40 °C and 0.1 MPa, and desorption at 110–120 °C and 0.2 MPa. The fed flue gases have an average CO_2 concentration in the range 2–14%, depending on the fossil-C used in the power plant for the production of electricity. In order to improve the environmental impact of the process and reduce costs, a variety of other "*sorbents*" have been tested that may have a good uptake capacity and selectivity toward other gases and reduce the energy of release of CO_2 [3].

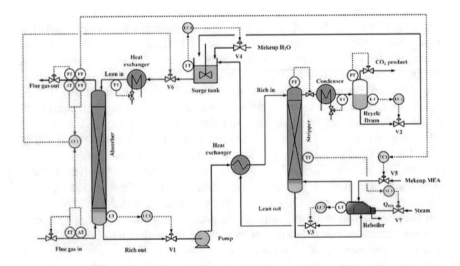

Fig. 6.1 A monoethanolamine plant for the separation (*uptake*, left tower and *release*, right tower) of CO_2 from flue gas. The separation efficiency ranges from 90 to 92.5%. Reproduced from Ref. [2b] (CC BY 4.0)

In addition to abovementioned metal oxides and amines, adsorbents such as C-based materials, pillared clays, molecular sieves, zeolites, silica gel, carbon nanotubes, and metal organic framework (MOFs) have been investigated, which often do not show the required selectivity in the separation from N$_2$, CH$_4$, and other gases. In general, chemical interactions result much more selective than physical processes in the separation of gases, but demand more energy. Amine-functionalized solid sorbents known since early 1990s [4] and the mechanism of which is known since early 2000s [5] have attracted great attention for CO$_2$ capture from flue gases and the atmosphere, as such materials have a lower heat capacity compared to water and, thus, demand a lower parasitic energy for uptake and release. Noteworthy, such new absorbers try to combine the selectivity of amines with the scarce volatility of solids. An "*ideal sorbent*," (Scheme 6.1) should not be easily degraded nor be volatile, so as to avoid atmosphere pollution, (see DI6.1) while showing low parasitic energy.

In fact, a major drawback of MEA (despite it finds application in operating plants since a few years) is its volatility that causes losses to the atmosphere, and its relatively low resistance to oxidants (present in flue gases) that cause loss and formation of derivatives, the impact of which on humans deserves still further deep investigation.

Polyetheneimine (PEI) is a polymer with alternate ethene moiety (–CH$_2$CH$_2$–) and imine moiety =NH. With respect to the amino moiety (–NH$_2$) of MEA, the imine group (=NH) has a lower capacity of capturing CO$_2$. Because of the abundance of N-centers in a more condensed form, PEI-based sorbents have been considered an interesting candidate for CO$_2$ capture from both point-concentrated CO$_2$ sources (power and industrial plants) and the atmosphere. As already mentioned, a point to note is that flue gases from power plants contain oxidants and acid species (O$_2$, NOx, SOy) which can degrade the materials used for separation.

Selected industrial streams (*ca.* 3500 Mt/y, Table 6.1) may contain less pollutants than flue gases from power stations and are more suited for CO$_2$ capture. Moreover, industrial sites are better suited to host a capture-conversion plant than power stations. Therefore, capture of CO$_2$ from industrial plants would be double beneficial. Fermentation plants produce quite pure CO$_2$ accompanied by water or

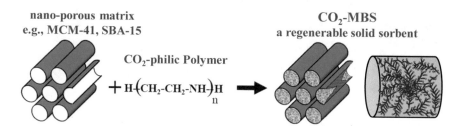

nano-porous matrix e.g., **MCM-41, SBA-15**   **CO$_2$-MBS** a regenerable solid sorbent

CO$_2$-philic Polymer

$+ \ \mathrm{H}\!\!\left(\!\mathrm{CH_2\text{-}CH_2\text{-}NH}\!\right)_{\!n}\!\mathrm{H}$

Scheme 6.1 Concept of "molecular basket" sorbents (MBS) for CO$_2$ capture. Reproduced by permission from Ref. [3b], Copyright 2009

Table 6.1 Emission of CO_2 from selected industrial processes

Industrial sector	Mt_{CO2}/y produced
Oil refineries	850–900
Petrochemical processes	155–300
LNG sweetening	25–30
Ethene oxide	10–15
Ammonia	160
Fermentation	>200
Iron and steel	ca. 900
Cement	>1 000

ethanol vapors which are easy to separate. Emissions from cement manufacture contain solid particles that must be separated before contacting the gas stream with the absorber in order to avoid the deactivation of the surface of the latter.

As reported in Table 6.2, major issues with the capture technology are the CAPEX and OPEX costs. The operational cost is still quite high due to the use of electric energy for the two steps of the separation process [6] and are believed to be in the range 50–120 US\$/tCO$_2$ depending on the quality of the stream of CO_2. Cleaner is the stream and higher the CO_2 concentration, lower will be the cost. Energy requirement for MEA plant operation is estimated to be in the range of 3.5–3.8 GJ/tCO$_2$, which must be compared to the energy consumption in case of a reversible process (no parasite losses) of 0.06 GJ/tCO$_2$ for IGCC and 0.7 GJ/tCO$_2$ for post-combustion capture or to estimated 0.96–1.24 GJ/t$_{CO2}$ for aqueous alkaline systems [7]. At BASF, a new process (OASE) has been developed [8a] that allows to save up to 35% of the energy necessary for CO_2 separation (2.7–3 GJ/tCO$_2$). JGC-Japan, INPEX, and BASF-DE have developed the HiPACT process (high-pressure acid gas capture technology) that improves the separation process and reduces costs by 25–35% [8b].

Research is going on for developing new absorbers that may minimize the overall energy cost, and are durable in time.

Table 6.2 Pros and cons of various separation technologies

Technology	Pros	Cons
Solid phases	Low loss	Energy demand
Liquid phases (LP), MEA	Mature	Loss, large volume
Membranes (M)	Low volume	Cost, lifetime
Combined (LP/M)	Efficiency	Volume, cost
Cryogenic	Low emission	Cost

Issues: CAPEX, OPEX, energetic costs (energy penalty: 20–40$^+$%)

6.3   Direct Capture from the Atmosphere

Recently, the direct capture from the atmosphere (DAC) has attracted a lot of interest as this technology decouples CO_2 capture from production sites and sources and makes the *capture* a *ubiquitous* technology. This is important whatever use we intend to do of CO_2: disposal or utilization. Although this implies a substantial work against entropy (455 kJ/kg$_{CO2}$), being the concentration of CO_2 quite low in the atmosphere (410 ppm or 0.041% v/v, more than two orders of magnitude lower than in flue gases) this must be considered as the future approach to CO_2 capture. The estimate of the actual economic cost of such operation is around 80–130 $/t$_{CO2}$ [9] even if much higher costs have been encountered in practice [10].

If one frames together the capture of CO_2 and its conversion into fuels (vide infra), then the fraction of energy which would be lost for CO_2 capture is important as it would reduce the net available energy of produced fuels. This aspect is dramatically more important in the case when fossil-C is used for powering the process, but can lose priority if perennial energy sources are used.

In DAC, large volumes of air are vehiculated over an absorber: either fans or pumps are used (Fig. 6.2). Then CO_2 is released and forwarded to the utilizer.

Several start-ups and companies have recently been established with the support of private and public funds for facing this exciting new challenge: to reduce the atmospheric load of CO_2. Despite plants have been developed to the demo scale, the energy balance is still quite obscure and it is difficult to establish the real economic and energetic cost of such interesting technology (see DI6.1).

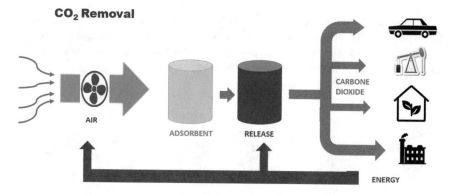

Fig. 6.2 Capture of CO_2 from the atmosphere

6.4   Recovery from Mobile Sources

An even more recent approach to cut the CO_2 emission is the capture from mobile sources. This technology can be applied to large carriers, such as trucks, buses, and boats while small vehicles are less suited for the lack of space where to allocate the absorbents. The basic idea is to place on the carrier a container filled with an absorber through which combustion gases are passed before the emission to the atmosphere occurs. CO_2 would be trapped in significant amount and fixed in the sorbent. The amount of CO_2 fixed would depend on the capacity of the absorber and on the kinetics of adsorption, as the contact time with the combustion gases would not be very long. Once exhausted, the container with the absorber would be deposed at specific stations for regeneration and CO_2 collection. CaO has been tested in such application, together with other materials. This technology is just in its infancy and it is unknown what is its real potential and cost.

6.5   Disposal of CO_2

As already mentioned, once captured from flue gases, CO_2 can either be disposed in natural fields (CCS) or used (CCU). The CCS concept had a large acceptance in the 1990s and 2000s and has experienced over 20 years of substantial investment in several countries all over the world, since 1996. Three major options were proposed: i. storage in spent natural-gas (NG) wells; ii. pumping into aquifers; and iii. deposition as solid CO_2 hydrate (Fig. 6.3, a white powder) in deep (<3000 m) water.

Despite CCS was presented as the technology that can capture and dispose all CO_2 formed by burning all the C-based fuels available on our planet, (Table 6.3) as for today less than 5 Mt_{CO2}/y are disposed, most of which in Enhanced Oil Recovery (EOR), a not real disposal technology.

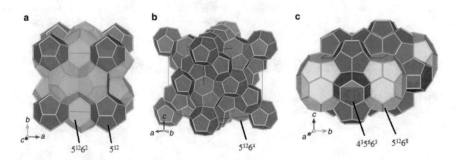

Fig. 6.3 Structure of solid hydrated CO_2: $CO_2 \cdot xH_2O$ ($x = 5.75–7.67$). Reprinted from Ref. [11], Copyright (2016), with permission from Springer

Table 6.3 Potential of storage of CO$_2$ in natural sites [12]

Storage option	Global capacity	
	Gt CO$_2$	% of emission to 2050
Depleted oil and gas field	920	45
Deep saline aquifers	400–10 000	20–500
Unmineable coal seams	>15	>1

The original enthusiasm has been frozen by our scarce knowledge of the persistence of CO$_2$ in disposal sites (after how long time it will come back to the atmosphere?) and the effect on living organisms if disposed in oceans. Spent NG wells can be considered a quite safe option because they already contained CO$_2$ and if they are correctly sealed after down-pumping CO$_2$, leakage should be minimized or even annulled. Conversely, aquifers are scarcely known for their leakage and hydrated CO$_2$ are quite unstable as unexpected warm water streams may convert the solid hydrate into gaseous CO$_2$ that can violently be released to the atmosphere causing serious damages. It is worth to recall here the violent emission of CO$_2$ that occurred on August 21, 1986 in Lake Nyos, Africa that caused 1 746 victims by asphyxia [13].

Moreover, waters rich in CO$_2$ have a low pH that can disturb the physiology of fishes. Therefore, as assessed at the recent summit on Innovative Technologies held in Houston in 2018, deep studies are necessary in order to gain knowledge of CCS effects before large-scale disposal may be implemented.

CCS requires a conscious political decision based on popular consensus. Several countries (USA, Japan, UK) have accepted CCS while others have rejected it (Germany, Austria, Denmark). Key issues are availability of sites, distance of the disposal site from the source, real economic cost, and real energetic cost (20–60% penalty on electric energy production including capture). Safe sites that can store CO$_2$ for long time are scarce and often far away from the point where CO$_2$ is generated.

An example which is worth to consider is the case of some European countries, such as Italy, that import natural gas from Siberia, some 5000+ km away. As shown in Table 6.4, for pumping methane from Siberia to Italy, a penalty of 9–18% of extracted methane is paid, and for sending back CO$_2$ (Fig. 6.4) the energy requested is higher because of the different physical state and viscosity of CO$_2$ with respect to CH$_4$ that require a higher number of pumping stations (Fig. 6.4). As a matter of fact, for sending back CO$_2$ to the place of extraction of methane, the energy penalty would be 36–72% of the energy produced from methane. Such energetic cost is not acceptable as it would imply a significant shortening of fossil resources. Therefore, a must for the implementation of CCS is that the storage site is close to the source of CO$_2$: a distance of 30 km implies a penalty of 20–25% of the electric energy produced, including capture. Such situation is not common, and logistics represents a serious drawback to an extensive exploitation of CCS. Three cases of disposal of CO$_2$ will be presented in different areas of our planet.

Table 6.4 Energy required in CH_4 and CO_2 pumping in a gas duct (St = Station)

Item	CH_4	CO_2
Physical status	Gas	Supercritical gas
Pressure	7.5 Mpa	12.0–18.0 MPa
Speed	20–30 km/h	10–20 km/h
Pumping St at each	80–160 km	40–80 km
Energy for pumping (ss)	0.15–0.30% v	0.6–1.2% emitted+
1000 km (12 St)	1.8–3.6%	7.2–14.4%
5000 km (60 St)	9.0–18%	36–72% (emitted/stored)

Fig. 6.4 Double gas duct for pumping methane from the extraction well to users and sending back CO_2

The Sleipner Project. An almost ideal situation occurs in Norway, at the offshore methane extraction wells. In 1991, the Norwegian Government introduced a CO_2 offshore tax with the aim of reducing CO_2 emissions (CO_2 separated from methane was vented). One of the two biggest CO_2 storage projects that European companies are involved in is the *Sleipner Project* in the North Sea. The projects involve stripping CO_2 from natural gas, bringing the CO_2 content down to the sale specification of the natural gas, and storing the excess in underground geological formations. Since 1996, 1 Mt of CO_2 per year have been separated from gas produced on the Sleipner Vest Field in the North Sea and stored in the Utsira Formation, a saline aquifer located 1 000 m below the seabed.

The Nagaoka field. The CO_2 injection started in July 2003 at Teikoku Oil Co. Ltd owned South Nagaoka gas field located in Nagaoka City, Niigata Prefecture, 200 km north of Tokyo and 12 km southeast from the coast of the Sea of Japan. The targeted *inland aquifer* for the CO_2 injection is a 12 m thick sandstone bed of early Pleistocene age, which lies about 1 100 m below the ground surface.

Purchased 99.9% pure CO_2 was injected in the supercritical state at the maximum rate of 40 t/d. As of June 27, 2004, the cumulative amount of injected CO_2 was 5 893 t, with a total amount of 10 000 t_{CO2} planned to be injected. The termination of the project was scheduled as 2005.

The Crick Bell Field. For the Bell Creek large-scale project, the PCOR Partnership is working with Denbury Onshore LLC to develop a robust, practical, and targeted monitoring verification, and accounting (MVA), risk management, and simulation project associated with commercial scale injection of CO_2 for the purpose of simultaneous CO_2 storage and CO_2 EOR. The PCOR Partnership region, which covers over 1.4 million square miles, emits approximately 510 million metric tons of CO_2 yearly from large stationary sources in the region. Overall, based on the current geological formations characterized, the PCOR Partnership region has the storage resource of 417 billion metric tons of CO_2 in saline formations, 2.9 billion metric tons in depleted oil fields, and 7.3 billion metric tons in unmineable coal seams, which is over four times the anticipated regional emissions over the next 100 years: assuming a static emission profile CCS has been considered as "the solution to CO_2 reduction" for its large capacity. After more than 20 years of investment (public and private), still many questions remain open and a lot of research is necessary in order to assess the real potential of CCS. The technology is "site specific": sites are not available everywhere within a reasonable distance to minimize energetic costs. Several countries do not accept such technology. Energy penalty varies with the distance and nature of the disposal site. Despite its great potential, the effective implementation is scarce: efforts are still going on.

One of the drawbacks is that CCS is not remunerative, it is a net cost: energetic (expanded fossil-C extraction) and economic. Only EOR is an economically convenient application.

6.6 Enhanced Oil Recovery (EOR) Produces an Economic Advantage

CO_2-EOR is not new. It was pioneered in the Permian Basin in West Texas in 1972, and the CO_2 for the first projects came from natural gas processing facilities. Afterward, natural CO_2, extracted from the ground, was used. CO_2-enhanced oil recovery (EOR) is considered to have the potential to significantly expand domestic oil supplies, increase job growth, and protect the environment. The CO_2-EOR technology has been used successfully in a number of oil plays (notably in West Texas, Wyoming, and Saskatchewan). The main components of EOR include the following:

(a) transportation of CO_2 to a mature oil field;
(b) injection of pressurized CO_2 into the field; and
(c) extraction of oil and separation from CO_2.

The cost of (a) may limit the application. It is obvious that the extraction of CO_2 from the ground for such operation means a net increase of the emission to the atmosphere, while if captured CO_2 is used, considering that part of it remains in the well, the net emission into the atmosphere would be reduced. Some technical barriers exist. Large projects for EOR are going on in Canada and USA (Fig. 6.5).

The additional extraction of oil pays for the use of CO_2. In fact, estimates suggest that recovery rates from existing reserves could be approximately doubled, while the application of EOR on a broad scale could raise domestic recoverable oil reserves in the United States by over 80 billion barrels [14b]. Similarly, there are claims [14c] that 4 470 fields, just over half of the known oil reservoirs in Alberta, Canada, are amenable to CO_2 injection for enhanced oil recovery. Moreover, [15] enhanced oil recovery applied in these reservoirs is believed to translate into an additional 165 billion barrels of oil recovered and over 1 Gt of CO_2 sequestration. It has been estimated [16] that at current oil and carbon prices and with current technology, approximately half of this capacity is economically viable.

Figure 6.6 shows the trend in EOR. It is evident that majority of projects are located in USA-Canada: the majority in the Permian Basin (Texas), and the rest in the USA-Canada joint project.

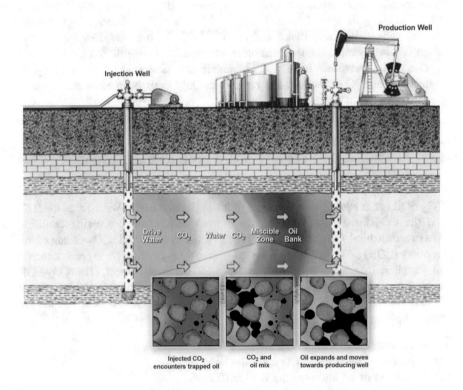

Fig. 6.5 The use of CO_2 in EOR. Supercritical CO_2 is injected to extract oil from rocks [14a]

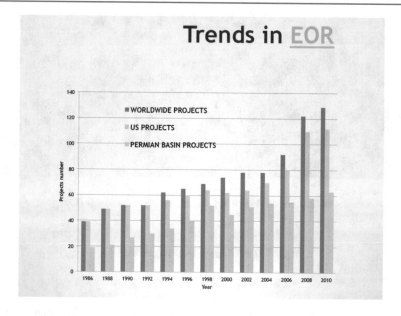

Fig. 6.6 Trend in CO_2 application in EOR

6.7 Fixation of CO_2 into Long-Lasting Inorganic Materials

A particular aspect of the utilization of CO_2 is its fixation into long-lasting materials such as inorganic carbonates. This option will be discussed in this chapter as it has some aspects similar to CCS: long time (centuries) permanence of CO_2 into the material. All other utilization options produce goods that will release CO_2 in a shorter time (from months for fuels, to years for chemicals, and few decades for polymeric materials).

The concept of storage of CO_2 into long living calcium or magnesium carbonates is referred as "*mineral carbonation*" (IPCC, 2005). The above carbonates are scarcely soluble in CO_2-free water and would represent a "*permanent*" storage of CO_2. CO_2-rich waters may attack the mineral as shown in Eq. 6.11.

$$Ca(Mg)CO_3 + H_2O + CO_2 \rightarrow Ca(Mg)(HCO_3)_2 \qquad (6.11)$$

Such solution could represent an alternative for those countries which do not have disposal sites, but requires the availability of basic oxides such as CaO and MgO. Such oxides are naturally available in rocks which are rich of basic components. Examples are olivine (Mg_2SiO_4) and wollastonite ($CaSiO_3$) (Fig. 6.7).

The conversion of such silicates into carbonates is a natural process known as "*natural weathering*" of silicate rocks (Eqs. 6.12a, 6.12b)

Fig. 6.7 Samples of forsterite olivine (left) [17a] and wollastonite (right) [17b]

$$Mg_2SiO_4 + CO_2 \rightarrow MgSiO_3 + MgCO_3 \tag{6.12a}$$

$$CaSiO_3 + CO_2 \rightarrow CaCO_3 + SiO_2 \tag{6.12b}$$

Several technologies have been developed to fasten such reactions that naturally occur on the timescale of centuries. Dissolution of rocks and precipitation of carbonates is an approach that, on the other side, might imply some environmental impact (see DI6.2). The use of basic industrial slag is an interesting approach as it would reach two targets: utilization of a waste and fixation of CO_2 into long-lasting materials. The production of mineral carbonates is a costly process with issues relevant to the cost of chemicals used and their recovery, production of liquid waste (cost of its treatment and fate of residues), end use of products, and environmental impact (if natural rocks are used). In order to avoid the use of chemicals and water, high-temperature gas-phase processes have been tested (see DI6.2).

An alternative approach to CO_2 storage into inorganic materials is the low-voltage electrochemical conversion of NaCl into $NaHCO_3$ or Na_2CO_3. Such process appears quite promising and economically and energetically viable.

The actual market of inorganic carbonates is around 200 Mt/y, with $CaCO_3$ leading with 113.9 Mt/y, followed by Na_2CO_3 (*ca.* 50 Mt/y), K_2CO_3, and other element carbonates (Ba, Sr, Cu, Fe).

Carbonates find application in several areas: $CaCO_3$ has large application as a filler in the paper and plastic industries, potassium carbonate is a component of soil additives, and several other carbonates find application in the production of special glasses. However, "*purity*" is one of the key requisites for an added value application of carbonates. Therefore, the quality of the starting materials is an issue. Industrial slag may contain elements that may affect human health and, thus, their separation is essential in order to avoid that they may enter the food chain, are swallowed/inhaled, or even get in touch with the person. For doing that, chemical technologies that may trap trace elements are necessary that increase the production cost (that should, instead, be as low as possible). Selective trapping of trace elements requires the use of designed complex (and costly) organic molecules that should be recovered after use, rising the complexity of the whole treatment. Even in industrial applications the purity can be a strict requirement. CaCO$_3$ can be produced in large amounts either from rocks or from slag, but often at low purity, so that it is used as source of bricks or as a filler in building roads, both being low value applications that may not pay for the cost of production of the material.

The use as filler in the paper industry is for sure a more profitable application that requires high quality, white, and finely dispersed CaCO$_3$. For producing such quality material, contaminants such as colored metal oxides must be eliminated: the purification process may be very costly if a complex matrix is used as an industrial slag or a multicomponent rock.

DI6.1: Capture of CO_2 from the Atmosphere or Direct Air Capture (DAC)

Revitalization of the atmosphere with capture of CO_2 is a key problem on board submarines and shuttles. Because of the low concentration of CO_2, absorbers which can tightly bind it, by forming strong chemical bonds, are necessary. Traditionally, lime (that combines NaOH and $Ca(OH)_2$) is used for CO_2 capture. The process, originally developed for the paper industry by Kraft (1884) for the extraction of cellulose, has been adapted to capture CO_2. It is based on a set of four reactions (Eqs. 6.13–6.16) which are as follows:

$$\text{Adsorber: } 2NaOH + CO_2 \rightarrow Na_2CO_3 + H_2O \quad \Delta H^\circ = -109.4 \text{ kJ mol}^{-1} \quad (6.13)$$

$$\text{Causticizer: } Na_2CO_3 + Ca(OH)_2 \rightarrow 2NaOH + CaCO_3 \quad \Delta H^\circ = -5.3 \text{ kJ mol}^{-1}$$
$$(6.14)$$

$$\text{Calciner: } CaCO_3 \rightarrow CaO + CO_2 \quad \Delta H^\circ = 179.2 \text{ kJ mol}^{-1} \quad (6.15)$$

$$\text{Slaker: } CaO + H_2O \rightarrow Ca(OH)_2 \quad \Delta H^\circ = -64.5 \text{ kJ mol}^{-1} \quad (6.16)$$

A sodium hydroxide water solution (1 M, or 1 mol/L) allows for effective binding of CO_2 and the formed carbonate (Eq. 6.13) is soluble in water (28.14 $g/100g_{H2O}$), differently from $CaCO_3$. Through reaction of the solution of sodium carbonate with calcium hydroxide (Eq. 6.14), calcium carbonate precipitates and can be filtered off, while sodium hydroxide is regenerated in solution, such process is known as causticization and has a calculated efficiency >96%. The calcination of $CaCO_3$ (Eq. 6.15) produces lime (CaO) and carbon dioxide, which can be collected and stored in cylinders. CaO is hydrated in a slaker (Eq. 6.16) and reused.

Besides sodium hydroxide, even potassium hydroxide (KOH) can be effectively used that has a higher cost. Noteworthy, on board shuttles, in order to reduce weight, LiOH is used. One mol of Li_2CO_3 has a mass of 82 g *versus* 106 g for Na_2CO_3. Therefore, per each mol of CO_2 fixed there is a weight reduction of 24 g. As a man emits ca. 900 g_{CO2}/d, the use of Li allows reduction of load of 491 g/d that means a lot on board a shuttle.

Since 1990s, when both the European Space Agency (ESA) and NASA launched a program to further develop the technology for capturing CO_2 on board of shuttles in view of deep space exploration (ultimate goal: sending man on Mars) [18], DAC has had a great push and has been further developed, but in a different context: use the technology as a means to remove CO_2 or other greenhouse gases from the Earth's atmosphere with the ultimate scope of fighting climate change. Several technologies are under scrutiny to this end (NaOH, $Ca(OH)_2$, membranes, and other absorbents) and several companies have already started marketing DAC as a climate solution, despite uncertainty existing about the real benefits and energetic and economic costs.

Searching the various investment programs one can find quite interesting information. (For the reader convenience we report the rate of change of currencies as for December 2020: 1 CAD\$ = 0.74 US\$; 1 EUR = 1.22 US\$; 1 £ = 1.34 US\$.)

In 2018, the Canadian start-up *Carbon Engineering Ltd.* claimed that DAC may be possible for US\$100/t_{CO2}. The process is based on capture by CaO, as reported above, and this will require heating at 900 °C for CO_2 release from $CaCO_3$. The current price of production by *Climeworks* is *ca.* 600 US\$/t_{CO2} with a perspective cost of 100 US\$/t_{CO2} by the end of the decade. Obviously, such energy-intensive technology requires that energy is provided by a cheap, available, and high-density non-fossil-C source. The *Climeworks* plant in Hinwil, Switzerland, needs 1 800–2 500 kWh of thermal energy and ~ 600 kWh of electricity to capture 1t of CO_2. One *Climeworks* collector can capture 50 t_{CO2}/y. If one would capture 5% of the total emission of CO_2 that amounts at 37 100 Mt/y [19], it means that for capturing 1 855 Mt/y of CO_2, at least 37.1 million collectors would be necessary. Such figures are impressive. Together with the energy required, the uncertainty on OPEX costs and environmental effects of waste materials, are on the list of problems to be solved for an effective use of such technology.

As for today, 10 companies worldwide located mainly in USA, Canada, and Europe (the majority are start-up and university Spin-off) plus a spin-off of the European Space Agency operate in this sector with the support of public (mainly in Europe) and private funding. According to the registered projects and the literature, some of them have raised substantial money during last few years: *Carbon Engineering* (Canada) > 80 MCAD\$, *Global Thermostat* (USA) > 70 MUS\$, *Climeworks* (ETH, Zurich, Switzerland) > 50 M€, and *Sunfire* (Germany, EU) 25 M€. Most of the projects aim at combining CO_2 capture and reduction to fuels (liquid or gaseous) by reaction with hydrogen produced by electrolysis of water using solar energy. The EU Horizon 2020 Program has funded the *SUNtoLIQUID* project that has established at the IMDEA Energy Institute (Madrid, Spain) a demo plant for making liquid hydrocarbons. Twenty-seven partners participated in The EU Project *STORE&GO* and aim at comparing three different technologies for converting captured CO_2 into methane. Even in this case hydrogen comes from water through PV electrolysis. *Soletair Power* funded by VTT-Helsinki (FI) similarly has set a plant for producing liquid fuels. The *Synhelion* project in Zurich uses the solar power concentrators for high-temperature CO_2 and water splitting to produce syngas and then liquid hydrocarbons by FT.

The *Sunfire GmbH* in Germany has set a demo plant for high-temperature electrolysis of CO_2 and water to afford syngas used for making liquid hydrocarbons.

Recently, DAC has been combined with CCS to assess the potential of CO_2 disposal (*CarbFix* Project). The *Rotterdam Jet Fuels* and the *Prometheus Fuels Projects* target the production of jet fuels. *Synhelion and ENI* target the production of methanol. *Silicon Kingdom Holding* is developing "*Mechanical Trees*" based on the use of sorbents (disks) that uptake CO_2 from dry air (20 min exposure at up to 10 m altitude). Once the sorbent is saturated the disk is lowered and CO_2 collected either by thermal or solvent process. The target is to collect up to 2.5 t_{CO2} per year with one "tree."

All above projects aim at developing technological solutions that may reduce the economic and energetic cost of DAC. A cost of 600 US\$/t_{CO2} is not economically acceptable and the technology will not reach commercialization. Recovery of CO_2 from more concentrated sources (point sources such as power plants and industry, for which the concentration of CO_2 ranges from 4 to 90+%) is much more convenient. Moreover, the application of CO_2 must be further developed and this requires investment in research. Data reported above say that there is a great interest in converting CO_2 and water into fuels, liquid, or gases. This attitude is fully at the hearth of the cyclic economy of carbon. In previous chapters, we have discussed the issue of "rate of formation/rate of conversion" of CO_2. We have shown that biological fixation of CO_2 is some 1 000–10 000 times slower than its formation rate. Chemical or catalytic conversion of CO_2 can be faster than the biological process. Will they be able to convert CO_2 into useful products at the same rate it is formed? This is a big challenge.

DI6.2: Mineral Carbonation: Great Expectations and Bottlenecks. The Potential Environmental Impact and Economic Cost

Mineral Carbonation (MC) is a scalable technology for long-term CO_2 sequestration, with the capacity to match the amount of CO_2 emitted from energy generation and industrial activities. Minerals, incineration ash, concrete, and industrial slag are sinks for anthropogenic CO_2 being a source of basic metal oxides, which can react with CO_2 to form inorganic carbonates and bicarbonates, thermodynamically stable and relatively inert at ambient conditions. If one looks at the abundance/availability of magnesium and calcium atoms on Earth (Fig. 6.8) [20a], can find that it is far exceeding the total amount of carbon atoms, but the rates of reaction to form (hydrogen)carbonates in Nature are too slow compared to the current rate of formation of CO_2.

Figure 6.9 is important for understanding the chemistry of CO_2 in carbonate mineralization.

CO_2 is sparingly soluble in water (0.027 g/L at 25 °C and 1 atm pressure of CO_2) [21]. It is worth to recall that the solubility refers to the *"free CO_2"* present in water and not to its hydrated forms that are much more abundant (see Fig. 6.9). Once dissolved in water, CO_2 is hydrated ($H_2O.CO_2$) and then forms the diprotic carbonic acid (H_2CO_3), a labile species which exists only in solution and cannot be isolated as it rapidly gets converted into H_2O and CO_2 (reverse of reaction 6.17a).

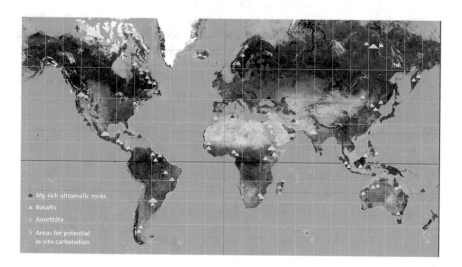

Mg-rich ultramafic rocks
Basalts
Anorthite
Areas for potential in-situ carbonation

Fig. 6.8 Red dots show regions where *ultramafic* (high content of *ma*gnesium and *ferric* materials) minerals are known to be present, worldwide. Their amount is estimated at *ca.* 90 000 Gt with the potential of storing *ca.* 22 000 Gt of CO_2. For comparison, the total amount of CO_2 from fossil-C is calculated to be *ca.* 10 000 Gt [20b]. Reproduced by permission from Ref. [20a] with permission of RSC

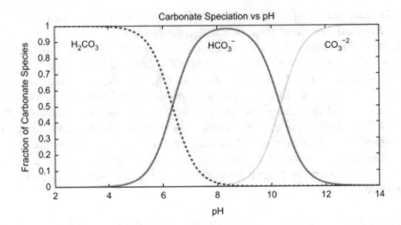

Fig. 6.9 Abundance of CO_2-derived species (H_2CO_3, HCO_3^-, CO_3^{2-}) as function of pH. Reprinted by permission from Ref. [21b], Elsevier, Copyright 2014

Carbonic acid, a weak acid, can undergo a double proton transfer to water affording the hydrogencarbonate anion (HCO_3^-) and the carbonate (CO_3^{2-}). The pH of the solution determines which species are dominating in solution. It is evident from Fig. 6.9 that below pH = 8 the carbonate species is practically converted into hydrogencarbonate or even hydrated CO_2 (H_2CO_3).

Often, the pH is not determined by dissolved CO_2, but from other adventitious stronger acid species such as organic acids formed upon degradation of biomass or sulfuric and nitric acid carried by acid rain. In this case, the distribution of carbonic species is determined by the external species. The contact of silicate rocks with acidic water may cause aggression with conversion of the silicate (Eqs. 6.17b–6.17d). Another reaction that must be considered is the carbonation of hydroxides (Eq. 6.17e), which occurs spontaneously.

$$H_2O + CO_2 \rightarrow H_2O \cdot CO_2 \rightarrow H_2CO_3 \qquad (6.17a)$$

$$CaSiO_3 + 2H_2CO_3 \rightarrow Ca(HCO_3)_2 + SiO_2 + CO_2 \qquad (6.17b)$$

$$Ca(HCO_3)_2 \rightarrow CaCO_3 + H_2O + CO_2 \qquad (6.17c)$$

$$CaSiO_3 + CO_2 \rightarrow CaCO_3 + SiO_2 \qquad (6.17d)$$

$$Ca(OH)_2 + CO_2 \rightarrow CaCO_3 + H_2O \qquad (6.17e)$$

Equation 6.17d represents the overall "*weathering*" process which needs acid solution to occur.

The process has kinetic barriers due to the slow dissolution in water and the low concentration that drive the aggression on the rocks, even if the thermodynamics is favorable.

The conversion of CO_2 ($\Delta G° = -396$ kJ mol^{-1}) into the carbonate anion ($\Delta G° = -530$ kJ mol^{-1}) has a negative free energy of *ca.* 134 kJ mol^{-1}. This is the effect of the combination of a strongly negative enthalpy change and a negative entropy change, due to the conversion of a gaseous species (CO_2) into a condensed one. The carboxylation of metal oxides, and of calcium oxide in particular, is a spontaneous reaction that means that the standard free energy (Eq. 6.18) is negative.

$$CaO(s) + CO_2(g) \rightarrow CaCO_3(s) \quad \Delta G° = \Delta H° - T\Delta S° = -253 \text{ kJ mol}^{-1} \quad (6.18)$$

Therefore, the carboxylation of metal oxide would release energy that can be recovered (and this is a positive fact), but needs to be carried on in solution (and this is a bottleneck).

Based on such data, how is, thus, performed the enhanced mineral carbonation? Let us start with the description of the materials that can be used, either natural rocks or industrial residues, as mentioned.

Among *natural rocks*, "*olivine*, Mg_2SiO_4" and "*serpentine*, $Mg_3Si_2O_7$ or $Mg_3Si_2O_5(OH)_4$, its hydrated form" are potential sources of magnesium for mineral carbonation. Olivine is an orthosilicate (Mg salt of orthosilicic acid H_4SiO_4; by comparison, "*enstatite*, $MgSiO_3$" is the Mg salt of metasilicic acid H_2SiO_3) made of isolated tetrahedral $[SiO_4]^{4-}$ units and interstitial Mg^{2+} cations (Fig. 6.10a).

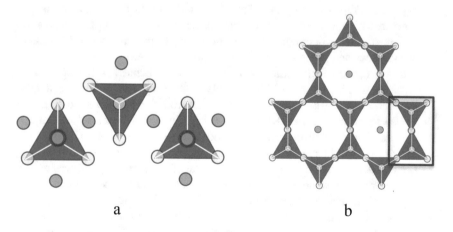

a   b

Fig. 6.10 On the left **a** are represented the single $SiO_4{}^{2-}$ tetrahedrals (Si = red, O = light green) and counterions M^{2+} (Mg^{2+} can be substituted by a number of different cations such as Fe^{2+}); on the right **b** is shown the lamellar phyllosilicate (phyllo from Greek Φυλλο = leaf) planar networks of polymeric interconnected $Si_2O_5{}^{2-}$ units (highlighted in the red rectangle in B), with the metal cations sitting in between two layers

Serpentine is a phyllosilicate made of parallel sheets of polymeric $(Si_2O_7^{6-})_n$ units interconnected by metal cations (Fig. 6.10b).

Olivine essentially is a volcanic material and is found in subsurface, where it stays out of the contact with atmosphere. Once it merges to surface it is changed into hydrated serpentine by weathering [22]. Being originated from another mineral, serpentine is said to be a *"metamorphic"* rock.

Serpentine exists as three different morphologies, namely, antigorite, lizardite, and chrysotile [23], the latter being better known as *"asbestos,"* now banned for its established cancerogenic (respiratory apparatus) properties when present in the air as tiny particles.

Abundant *industrial sources* of basic oxides include i. *steel slag*, ii. *waste concrete*, iii. *red mud*, and iv. *incineration bottom ash*. Other sources exist, which have a limited capacity.

i. A *basic slag* is formed during the smelting of iron ore, a process which requires lime to remove acid impurities (Si, Ti, and others oxides). A solid calcium silicate phase is formed, which floats on top of the molten iron and is skimmed off. This slag has an average CaO content of around 40–50 w/w%. The annual production of steel and iron slag is in the range 470–610 Mt [24], which corresponds to a CO_2 sequestration capacity of approximately 143–186 Mt/y. The calcium carbonate produced in this way can find application in buildings or as filling material in road construction.

ii. *Waste concrete*, generated in the demolition of buildings, is either disposed or recycled as aggregates in new construction projects. However, the fraction of waste concrete that is disposed of every year could be utilized as carbon sink for CO_2. Concrete contains significant amounts of calcium oxide, capable of capturing CO_2 as it settles and hardens. As already discussed above, the carbonation of concrete particles occurs on the surface, causing a gradual retardation of the penetration of CO_2 into the deeper layers where basicity accumulates and helps to prevent steel (iron) oxidation.

iii. *Bauxite*, the ore from which aluminum (Al) is obtained is treated with an aqueous solution of NaOH that dissolves the natural Al oxides as tetrahydroxoaluminate $(Al(OH)_4^-)$ that is separated from the residual solids and processed to extract Al_2O_3. The residual solids, known as *"red mud"* because of the presence of reddish iron (hydr)oxides, are strongly basic. The over 150 Mt/y of red mud have a CO_2 uptake capacity ranging from 0.03 to 0.08 t_{CO2}/$t_{red\ mud}$, depending on its quality, with a total of 4–12 Mt_{CO2}/y sequestered. As an apparently low amount, this beneficial action should be summed to the lower environmental impact caused by lowering the burden of the slag.

iv. *Incineration Bottom Ash* (IBA), formed in the combustion of coal [25] or Municipal Solid Waste (MSW), is rich of basic oxides that are potential binders of CO_2. Such IBA is processed to eliminate ferrous materials and, then, potential pollutants by treatment with lime. Residual lime increases the content of basic components that make this material useful for CO_2 sequestration. The combustion of Municipal Solid Waste (MSW) varies with countries and carbonation of IBA can locally represent the sequestration of a small fraction of emitted CO_2.

Table 6.5 presents the average Ca, Mg abundance in the materials discussed in this DI6.2, together with other components.

It is, thus, evident that the materials, either natural or man-made, that can sequestrate CO_2 are quite complex and, thus, require complex processes for CO_2 fixation, as reported in following paragraphs.

Serpentine is the best source of Mg, while wollastonite, steel slag, cement waste, and IBA are the most suited sources of Ca.

The processes for mineral carbonation can be categorized as (a) three-phase gas–liquid–solid direct process and (b) liquid phase chemically driven carbonation. Any combination of these approaches is also possible for increasing the conversion yield.

(a) The *"Pressure" process* works at a temperature of 100–185 °C and a pressure of 4–15 MPa, depending on the materials used (see Table 6.6) and consists of an injection of CO_2 and water vapor with other chemicals, into an autoclave containing the solid crushed and dried material.

Table 6.5 Average composition of natural (serpentine, wollastonite) and industrial materials that can supply *Ca* or *Mg oxides* for mineral carbonation. Figures represent w/w%

	Serpentine	Wollastonite	Steel slag	Cement waste	Incineration bottom ash
CaO	5	40–50	45–55	60–70	45–55
MgO	35–45	Minor	4.5	2	2.5
Al_2O_3	5	0.5	2.5	5	5
SiO_2	40	50	18	20	12
Fe_2O_3	10	Minor	2	2.5	10
Other components	Cr, Mn, Ni, others	K, Mn, Ti, P, others	Mn, Ti, S, others	K, P, Ti, Mn, others	Na, P, S, Cl, K, others

Table 6.6 Response of various materials to the *direct carbonation* in a three-phase process

Material	T (°C)	P_{CO2} (MPa)	Other additives	Conversion yield %
Olivine	185	15	NaCl, 1 M, $NaHCO_3$, 0.64 M	ca. 80–92%
Wollastonite	100	4	Neat	ca. 80+%
Serpentine	155	11.5	NaCl, 1 M, $NaHCO_3$, 0.64 M	ca. 80+%
Lizardite	150	10–15	NaCl, 1 M, $NaHCO_3$, 0.64 M	ca. 36%

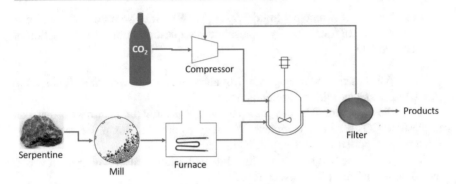

Fig. 6.11 Schematic view of a plant for the direct carbonation of natural minerals

Figure 6.11 shows the scheme of a plant for the direct carbonation of natural minerals.

The main drawback of such process is that the two main processes require opposite conditions: the aggression on the mineral is facilitated by acid conditions while the formation and precipitation of the carbonate require basic conditions. These opposite requirements make that the "one-pot" process is not highly efficient, depending on the nature of the starting material. An attempt to improve conversion yield has been made with revisiting the process and the concept of the plant. At NETL, an alternative process was developed that is based on a two-step process [26]. The overall reaction is depicted in Eq. 6.19.

$$Mg_3Si_2O_7(3MgO.2SiO_2) + 3CO_2 \rightarrow 3MgCO_3 + 2SiO_2 \qquad (6.19)$$

However, the main difference between the one-pot and the two-step process is that the separation of the carbonation–precipitation reactions makes possible their optimization in terms of temperature, pressure, and overall conversion yield. A $P_{CO2} = 0.5$ MPa was enough to carry the reaction at 90 °C with an almost quantitative yield of available $MgCO_3$.

Chemical carbonation is based on a different approach as it makes use of chemical energy to dissolve the mineral rock. This allows to run the reaction at ambient conditions avoiding high temperature and pressure, but introduces other negativities. Hydrated serpentine [$Mg_2Si_2O_5(OH)_4$] reacts with a water solution of a strong acid such as HCl to solubilize magnesium as $MgCl_2$, more than the dried form, leaving solid silica that can be separated. Bringing magnesium in solution has the positive effect of accelerating the reaction as reactions in solution are faster than in the solid state and, additionally, make available all Mg or Ca centers that is not the case in the solid state. In solution, the sequence of reactions is as depicted in Eqs. 6.20–6.22.

$$Mg_3Si_2O_5(OH)_4 + 6HCl \rightarrow 3MgCl_2 + 2SiO_2 + 5H_2O \qquad (6.20)$$

$$CO_2 + 2NaOH \rightarrow Na_2CO_3 + H_2O \qquad (6.21)$$

$$MgCl_2 + Na_2CO_3 \rightarrow MgCO_3 + 2\,NaCl \qquad (6.22)$$

As shown in Eqs. 6.20–6.22, the added chemicals HCl and NaOH are converted into NaCl that cannot be recycled. SiO_2 formed in Eq. 6.20 can easily be filtered off, and so $MgCO_3$ formed in Eq. 6.22. For keeping low the cost of the process and the environmental impact, NaCl should be converted back into HCl and NaOH: this is the energy demanding part of the process.

A process that recycles NaCl has been developed at ICES-Singapore (Fig. 6.12).

In this process, $MgCl_2$, less stable than NaCl by 59 kJ mol^{-1}, is thermally decomposed with water to afford MgO, sent to carbonation, and HCl that is dissolved in water and recycled.

HCl is not the only acid that can accelerate the dissolution of rocks and, thus, a diversity of processes have been developed all around the world with the target of finding the less energy demanding system for a closed cycle that would avoid the production of waste. Figure 6.13 shows the ammonium sulfate process developed at ABO-Finland that recycles ammonium sulfate used for dissolving the rocky material serpentine.

Other carbonation strategies have been used such as the use of enzymes or coupling with other processes for equilibrium shift. *Carbonic anhydrase,* present in several living organisms, is an enzyme that promotes the elimination of CO_2 formed at cellular level by converting it into HCO_3^- and dehydrating such species very fast at the lung level so that CO_2 is eliminated with expiration, avoiding its accumulation that can affect health of an organism. Such enzymes [28] and its inorganic mimics based on Ni complexes have been used with alternate success and open questions about cost, impact, and effectiveness.

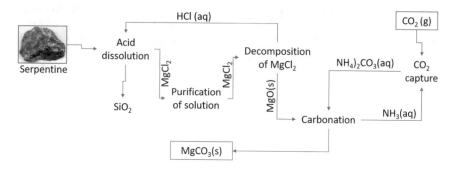

Fig. 6.12 Hydrochloric acid recycling in the ICES process [27]

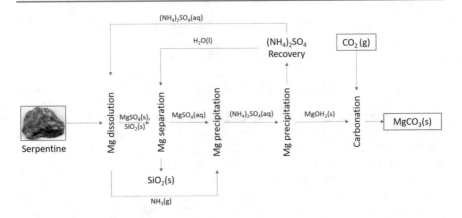

Fig. 6.13 The ammonium sulfate process developed at ABO-Finland

It is obvious that the *energetic balance* in all these processes has a fundamental role as the balance of CO_2 cannot be negative in the sense that does not make sense to emit more CO_2 than is sequestered into the carbonates. Therefore, the *CO_2 penalty*, expressed as the *ratio of emitted CO_2 over sequestered CO_2*, has a fundamental role: it must be lesser than 1, the lower it is better is the process. As the product in all processes is the carbonate, the comparison of processes can be done considering the energy input that depends on the technical aspects. Avoiding pressure and lowering the temperature are fundamental parameters together with the reduction of energy in recycling of materials.

Besides energy parameters, Investment (CAPEX) and Operative costs (OPEX) have a role in decision-making besides the quality of the product that will determine its value and use.

A point of importance is that the cost of the technology, whatever it is, depends on the site where the plant is located. As shown in Fig. 6.14, the parameters on which the process depends are numerous: the target is the optimization of all of them in order to minimize energetic and economic costs and maximize profits.

All products and by-products have a market, for example, $MgCO_3$ has a market price in the range 500–2 500 US\$/t according to its purity and particle size, while $CaCO_3$ has a much lower price, up to 300–500 US\$/t, but as shown in Table 6.4, Ca is much more abundant than Mg.

Fig. 6.14 Schematic view of the phases of mineral carbonation, products, and by-products obtained

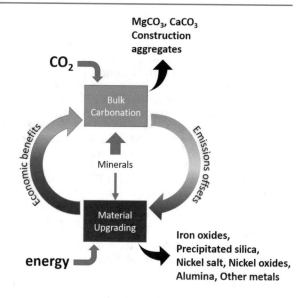

References

1. (a) Aresta M, Forti G (eds) (1987) Carbon dioxide as a source of carbon: chemical and biochemical uses, Reidel Publ., NATO—ASI: Dordrecht; (b) Aresta M (ed) (2003) Carbon dioxide recovery and utilization. Kluwer Publ; (c) Aresta M (ed) (2010) Carbon dioxide as chemical feedstock. Wiley VCH; (d) Aresta M, Dibenedetto A, Angelini A (2013) The changing paradigm in CO_2 utilization. J CO_2 Util 3–4:65–73
2. (a) Aresta M, Dibenedetto A, Quaranta E (2016) Reaction mechanisms in carbon dioxide conversion. Springer Publ; (b) Chen Y-H, Shen M-T, Chang H, Ho C-D (2019) Control of solvent-based post-combustion carbon capture process with optimal operation conditions. Processes 7:366
3. (a) Wang XX, Ma XL, Song CS, Locke DR, Siefert S, Winans RE, Mollmer J, Lange M, Moller A, Glaser R (2013) Molecular basket sorbents polyethyleneimmine with SB15 for carbon dioxide capture. Microporous Mesoporous Mater 169:103–111; (b) Wang XX, Song CS (2019) Capture of CO_2 from concentrated sources and the atmosphere. In: Aresta M, Karimi I, Kawi S (eds) An economy based on CO_2 and water, Chap 2. Springer
4. Tsuda T, Fujiwara T (1992) Polyethyleneimine and macrocyclic polyamine silica gels acting as carbon dioxide absorbents. J Chem Soc Chem Commun 1659–1661
5. (a) Dibenedetto A, Aresta M, Fragale C, Narracci M (2002) Reaction of silylalkylmono-and silylalkyldi-amines with carbon dioxide: evidence of formation of inter-and intra-molecular ammonium carbamates and their conversion into organic carbamates of industrial interest under carbon dioxide catalysis. Green Chem 4(5):439–443; (b) Dibenedetto A, Pastore C, Fragale C, Aresta M (2008) Hybrid materials for CO_2 uptake from simulated flue gases: xerogels containing diamines. ChemSusChem 1(8–9):742–745
6. McKinsey&Company, Inc. (2008) Carbon capture & storage: assessing the economics
7. Chan HXM, Yap EH, Ho JH (2013) Overview of axial compression for DAC of CO_2. Adv Mater Res 744:392–395
8. (a) www.oase.basf.com; (b) Kumagai T, Tanaka K, Fujimura Y, Ono T, Ito F, Katz T, Spuhl O, Tan A (2011) HiPACT–advanced CO_2 capture technology for green natural gas exploration. Energy Procedia 4:125–132
9. Holmes G, Keith DW (2012) An air–liquid contactor for large-scale capture of CO_2 from air. Phil Trans R Soc A 370:4380–4403
10. http://www.geoengineeringmonitor.org/2018/05/direct-air-capture/

11. See Chapter 10 In: Aresta M, Dibenedetto A, Quaranta E (2016) Reaction mechanisms in carbon dioxide conversion. Springer Publ

12. https://www.sciencealert.com/how-this-small-lake-in-africa-once-killed-1-700-people-overnight-and-we-still-don-t-know-why

13. IEA Greenhouse Gase R&D Programme (2001)

14. (a) https://www.energy.gov/fe/science-innovation/oil-gas-research/enhanced-oil-recovery; (b) Evaluating the potential for "Game changer" improvements in oil recovery efficiency from CO_2 enhanced oil recovery. Advanced Resources International, Prepared for U.S. Department of Energy Office of Fossil Energy—Office of Oil and Natural Gas. ARI (2006); (c) Shaw JC, Bachu S (2003) Evaluation of the CO_2 sequestration capacity in Alberta's oil and gas reservoirs at depletion and the effect of underlying aquifers. J Can Petroleum Technol 42(9):51–61

15. Babadagli T (2006) Optimization of CO_2 injection for sequestration/enhanced oil recovery and current status in Canada. In: Lombardi S, Altunina LK, Beaubien SE (eds) Advances in the geological storage of carbon dioxide: international approaches to reduce anthropogenic greenhouse gas emissions. Springer, Dordrecht, The Netherlands, pp 261–270

16. Snyder S, Anderson I, Keith D, Sendall K, Youzwa P (2008) Canada's fossil energy future: the way forward on carbon capture and storage. Report by the ecoENERGY Carbon Capture and Storage Task Force to the Minister of Alberta Energy and the Minister of Natural Resources Canada

17. (a) Geology Science (2020) Olivine [online]. https://geologyscience.com/minerals/olivine/. Accessed 3 July 2020; (b) Geology Science (2020) Wollastonite [online]. https://geologyscience.com/minerals/wollastonite/. Accessed 4 July 2020

18. Kay R (1998) International Space Station (ISS) Carbon Dioxide Removal Assembly (CDRA) protoflight performance testing. In: International conference on environmental systems, SAE International, Danvers, Massachusetts

19. Muntean M, Guizzardi D, Schaaf E, Crippa M, Solazzo E, Olivier J, Vignati E (2018) Fossil CO_2 emissions of all world countries. In: Environment and climate change. Publications Office of the European Union. ISBN: 978-92-79-97240-9

20. (a) Sanna A, Uibu M, Caramanna G, Kuusik R, Maroto-Valer MM (2014) A review of mineral carbonation technologies to sequester CO_2. Chem Soc Rev 43:8049–8080; (b) Matter JM, Kelemen PB (2009) Permanent storage of carbon dioxide in geological reservoirs by mineral carbonation. Nat Geosci 2(12):837

21. (a) Rumble J (2019) CRC Handbook of chemistry; (b) Vandehey NT, O'Neil JP (2014) Capturing [^{11}C]CO_2 for use in aqueous applications. Appl Radiat Isotopes 90:74–78

22. Moody JB (1976) Serpentinization: a review. Lithos 9(2):125–138

23. (a) Rinaudo C, Gastaldi D, Belluso E (2003) Characterization of chrysotile, antigorite and lizardite by FT-Raman spectroscopy. Can Mineral 41(4):883–890; (b) Groppo C, Rinaudo C, Cairo S, Gastaldi D, Compagnoni R (2006) Micro-Raman spectroscopy for a quick and reliable identification of serpentine minerals from ultramafics. Eur J Mineral 18(3):319–329

24. Shi C (2004) Steel slag—its production, processing, characteristics, and cementitious properties. J Mat Civil Eng 16(3):230–236

25. Armesto L, Merino JL (1999) Characterization of some coal combustion solid residues. Fuel 78:613–618

26. (a) Geerlings JJC, Wesker E (2008) Process for sequestration of carbon dioxide by mineral carbonation. WO2008142017 (A2) (A3), EP2158158 (A2), CA2687618 (A1), AU2008253068 (A1), US20070261947 (A1), CN101679059 (A); Shell International Research Maatschappij B.V; (b) Geerlings JJC, Van Mossel GAF, Veen BCMIT (2007) Process for sequestration of carbon dioxide, WO2007071633A1

27. Yeo TY, Bu J (2019) In: Aresta M, Karimi I, Kawi S (2019) An economy based on CO_2 and water, Chap 4. Springer

28. Power IM, Harrison AL, Dipple GM (2016) Accelerating mineral carbonation using carbonic anhydrase. Environ Sci Technol 50(5):2610–2618

Properties of the Carbon Dioxide Molecule

7

Abstract

The basic aspects of the reactivity of carbon dioxide are featured, related to the electronic structure of the molecule. The phase diagram of CO_2 is also discussed for understanding how CO_2 exists in different conditions of temperature and pressure.

7.1 Electronic Properties of CO_2

The knowledge of the electronic (and physical) properties of CO_2 is fundamental for understanding the reactivity of CO_2, which means the behavior toward a variety of reagents and in diverse conditions, which will be discussed in following chapters. While we have tried to make a very basic approach to such complex subject, nevertheless, the arguments are quite specific and may not be of common understanding. We invite the reader (hoping there will be at least one!) to try to catch the essential message and take home the fundamentals for understanding the behavior of CO_2 in reactive systems. In the Dictionary of Terms in the Appendix, we have tried to "vulgarize" main terms and concepts trying to make them more accessible. Thanks for the effort you will do going through the following text!

7.1.1 The Ground State of Carbon Dioxide and Its Geometry

In its electronic ground state, the carbon dioxide molecule has a linear geometry (Fig. 7.1) and belongs to the point group $D_{\infty h}$. Both C-O bonds are equivalent with an equilibrium distance equal to 116.00 pm, as established by electron diffraction [1]. Both carbon–oxygen bonds are polar, due to the higher capacity of the O-atom with respect to carbon of attracting electrons (higher electronegativity) [2].

© Springer Nature Switzerland AG 2021
M. Aresta and A. Dibenedetto, *The Carbon Dioxide Revolution*,
https://doi.org/10.1007/978-3-030-59061-1_7

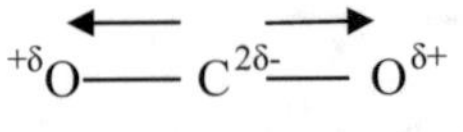

$$^{+\delta}O \overset{\longleftarrow}{\underline{}} C^{2\delta-} \overset{\longrightarrow}{\underline{}} O^{\delta+}$$

Fig. 7.1 Dipole moments in the carbon dioxide molecule

However, the two dipole vectors have equal magnitude and point in opposite directions, cancelling each other due to the linear geometry of the molecule that, thus, has no permanent electric dipole and is *apolar* (Fig. 7.1).

Anyway, the two dipoles originate a non-zero (-4.3×10^{-26} esu/cm^2) electric quadrupole [3a, g] that originates significant intermolecular interactions, which may account for the formation of neutral clusters $(CO_2)_n$ ($2 \leq n \leq 5$) observed via mass spectrometry [3b, d]. The intermolecular interaction can be reinforced if a molecule of CO_2 bears a positive or negative charge producing very stable charged aggregates $(CO_2)_n^{+(-)}$ ($2 \leq n \leq 10$), models for CO_2-solvated CO_2^+ radical cation or solvated CO_2^- radical anion [3c–e].

Figure 7.2 shows the energy diagram of the Molecular Orbitals-MO of the carbon dioxide molecule. It is worth to note that only the *"valence"* atomic orbitals (AO) are shown in Fig. 7.2 (the C, O 1s orbitals are, thus, not represented), and 2s AOs of the oxygen atoms are not considered in bond formation because they lie too low in energy (-37.6 eV): such orbitals ($3\sigma_g$ and $2\sigma_u$) will be located on the two O-atoms. Conversely, $2s_C$ (-19.4 eV) and $2p_C$ (-10.7 eV) are combined with 2p of the two O-atoms to form molecular orbitals. Overall, we have four σ-MOs marked in Fig. 7.2 as $4\sigma_g$, $5\sigma_g$, $3\sigma_u$, and $4\sigma_u$ formed by the combination of the $2s_C$ and $2p_{z,C}$ with two $2p_{z,O}$ (one from each O-atom, z is considered the molecular axis) and

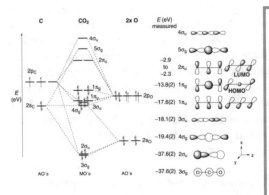

The *electronic configuration* of the linear ground state, $^1\Sigma_g^+$, of the 16e⁻ CO_2 molecule (the carbon atom has four valence electrons, while the oxygen atom has 6 valence electrons, altogether makes $(2\times6)+4=16$) is reported below. (See the graphical representation in Fig. 7.2)

$^1\Sigma_g^+$ (ground state): $1\sigma_u^2$ $1\sigma_g^2$ (-541.1 eV) $2\sigma_g^2$ (-297.5 eV) $3\sigma_g^2$ (-37.6 eV) $2\sigma_u^2$ (-37.6 eV) $4\sigma_g^2$ (-19.4 eV) $3\sigma_u^2$ (-18.1 eV) $1\pi_u^4$ (-17.6 eV) $1\pi_g^4$(-13.8 eV)

The values in parenthesis provide an approximate estimate of CO_2 molecular orbitals (MO) energy evaluated by measuring the ESCA ionization energies for the molecule [4a-c]. ESCA cannot split the energy of $3\sigma_g$ and $2\sigma_u$ orbitals. Nevertheless, calculations have demonstrated that $3\sigma_g$ is lower in energy. It is worth to recall that even if the value of the energy of orbitals may be affected by the method used in calculation, nevertheless, the relative order should not change with the method used.

Fig. 7.2 Energy diagram for the valence molecular orbitals of ground-state CO_2. On the *left* are represented the energy levels ("atomic orbitals") of the C-atom in its ground state, on the *right* those of the O-atom, and in the middle the molecular energy levels ("molecular orbitals") for the CO_2 molecule, generated by linear combination of atomic orbitals of the two O and C atoms. To each "orbital," either atomic or molecular, two electrons can be associated, having "antiparallel spin" or sense of rotation around its own axis (one spins to the right and the other to the left). Reprinted from Ref. [4d], Copyright (2017), with permission from Elsevier

three doubly degenerate (having the same energy) π-orbitals denoted as $1\pi_u$, $1\pi_g$, and $2\pi_u$.

Schematic and simplified representations of the CO_2 molecular orbitals are shown in Fig. 7.3. Figure 7.4 illustrates qualitatively the variation of the MO energies with the change of OCO bond angle (diagram of Walsh [5]) from the linear (180°, $D_{\infty h}$) to the bent (90°, C_{2v}) configuration. It may be useful to remind that the most commonly found bent geometry is at ca. 133°, as in CO_2^-, in organic carboxylates and inorganic carbonates.

The 16 valence electrons of the CO_2 molecule occupy the orbitals from $3\sigma_g$ and $2\sigma_u$ up to $1\pi_g$, the Highest (in energy) Occupied Molecular Orbital (HOMO), located essentially on the O-atoms. The next in energy is the $2\pi_u$ that is the Lowest Unoccupied Molecular Orbital (LUMO). *HOMO* and *LUMO* are often known as "*frontier orbitals*" and are implied in electron transfer within the CO_2 molecule (excitation), or to the CO_2 molecule from an external donor (CO_2 acting as an acceptor or acid), or even from the CO_2 molecule to an external acceptor (CO_2 acting as a donor or base).

The doubly degenerate $1\pi_{ux}(1b_1)$ and $1\pi_{uy}(5a_1)$ orbitals are perpendicular to each other and to the molecular axis. They are, respectively, bonding combinations of $2p_x$ and $2p_y$ states on all the three atoms. In the linear configuration also, the doubly degenerate $1\pi_{gx}(1a_2)$ and $1\pi_{gy}(4b_2)$ molecular orbitals (antibonding combination of $2p_{xO}$ and $2p_{yO}$, respectively) are equivalent and reciprocally rotated by 90° about

Fig. 7.3 Schematic view of the CO_2 molecular orbitals: white and black indicate the positive and negative signs of the wave function. Reprinted from Ref. [5b], Copyright (2016), with permission from Elsevier

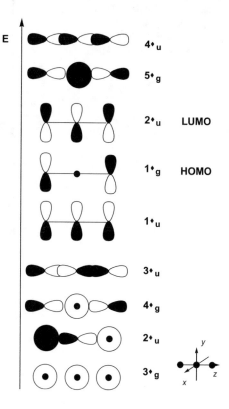

the molecular axis. They do not involve the C-atom (nodal plane) and behave more as "lone pairs" located on the O-atoms.

7.1.2 The Lowest Excited States of Carbon Dioxide

The energy of either HOMO or LUMO molecular orbitals may undergo significant change upon bending the molecule, as shown in Fig. 7.4. In general, the energy of the σ-orbitals either increases or is almost unaffected upon molecular bending, while the energy of the π-orbitals can either increase or decrease or even remain almost unchanged, depending on the direction of bending. In fact, assuming the Z-axis as the molecular axis, bending can occur either in the X- or Y-direction; consequently, the two degenerate orbitals will behave differently whether they are in the direction of bending or perpendicular to it, with loss of degeneracy. Why such splitting is observed? The explanation is quite simple: upon bending in the right direction, electrons located on the two oxygen atoms may get closer,

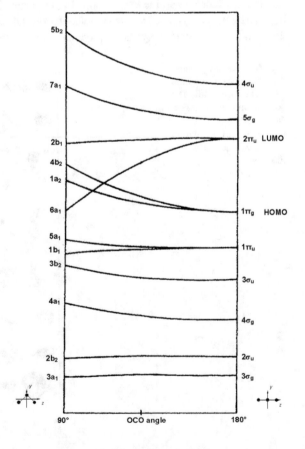

Fig. 7.4 Change of the orbital energy moving from the linear to the bent (90°) geometry. (Simplified Walsh Diagram), Reprinted from Ref. [5b], Copyright (2016), with permission from Elsevier

increasing their repulsion and the energy of the orbitals. The linear geometry is the one that minimizes electron repulsion and thus has the lower energy.

Figure 7.4 makes evident the peculiar character of the empty $2\pi_u(6a_1)$ MO. Upon bending its energy largely decreases. However, any electron transfer to such orbital, whatever is the origin of the electron, will reduce the energy upon bending of CO_2. This is true when CO_2 is excited (internal e^--transfer), when a single electron is transferred to CO_2 to form the radical anion CO_2^-, or when two electrons are used in bonding CO_2 at the C-atom. The internal electron transfer corresponds to the "excitation of CO_2"; in addition to bending, an elongation of the C–O bond from original 116 pm to 124–126 pm is observed, while the angle moves from original 180° to 122°–148°, according to the state.

7.1.3   Main Features of Carbon Dioxide Reactivity

CO_2 exhibits an amphoteric character, the oxygen atoms acting as Lewis bases and the carbon atom as a Lewis acid center (Fig. 7.5). However, carbon dioxide is a better acceptor than donor of electron density and, consequently, the reactivity of the molecule is dominated by the electrophilic character of carbon rather than by the weak nucleophilic properties of the oxygen atoms.

7.1.3.1   Carbon Dioxide as O-Nucleophile

A measure of the weak basic character of the oxygen atoms of CO_2 is provided by the value of its proton affinity 540.5 ± 2 kJ/mol [6a]. Such measured value is appreciably lower than those found for other O-containing molecules, such as H_2O, MeOH, Me_2O, H_2CO, MeCHO, Me_2CO, HCO_2H, and $MeCO_2H$ [6b, c].

Hydroxycarbonyl cation $HOCO^+$ (Eq. 7.1) is the simplest adduct in which CO_2 acts as a O-nucleophile. It is thought to be an important intermediate species in gas-phase reactions in interstellar clouds and space [7].

$$CO_2 + H^+ \rightarrow HOCO^+ \tag{7.1}$$

Submillimeter wave spectroscopy [8] and infrared spectroscopy studies [9] have shown that the interstellar lines observed in 1981 by Thaddeus et al. [10] in the 85 GHz region belong to $HOCO^+$. The equilibrium values for the OCO and COH

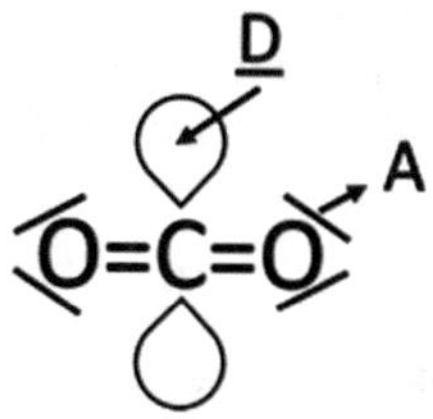

Fig. 7.5   Lewis acid–base properties of CO_2. **A** is a Lewis acid, **D** is a Lewis base

angles, as recently computed by Hammani et al. [11] using the coupled electron pair approximation (CEPA) method with a correlation consistent basis set, are, respectively, equal to 174.3° and 117.6°, while the O–C, C–OH, and O–H bond lengths are, respectively, of 112.4, 122.9, and 98.4 pm. Theoretical investigations agree on the fact that the OCO backbone of this molecule is slightly bent, with a *trans* configuration and the C–O and C–OH bond distances being, respectively, shorter and longer than what is observed in the non-protonated molecule [6b, c, 11]. The positive charge is mainly localized on the carbon and hydrogen atoms with the following repartition: C, +0.584; H, +0.350; O, +0.012; and (H)O, +0.054. The alternative C-protonation affords a cyclic dioxyril cation with a very weak O–O (1.733 Å) bond and a very high positive charge (+0.89) at the central CH unit.

Alternative to the protonation is the interaction with metal cations: several examples of O-coordinated CO_2 to metal centers are known. The available structural data show unambiguously that, whenever CO_2 behaves exclusively as an O-nucleophile, the coordinated CO_2 molecule essentially retains its original linear geometry or undergoes only slight distortion from linearity. This suggests that the interaction of CO_2 with an electrophilic center through one of the $1\pi_g$ lone pairs is not accompanied by any significant back-donation of electron density into the $2\pi_u$ orbitals of the heterocumulene, which, therefore, are left empty.

End-on (η^1-OCO) coordination mode has remained elusive for a long time. Nevertheless, this coordination mode is now well documented [12] and plays an important role in biological systems, such as photosynthetic CO_2 fixation by ribulose-1,5-biphosphate carboxylase-oxygenase (RuBisCO) [13a]. A crystallographic study on a deacetoxycephalosporin C synthase (DAOCS) mutant showed the presence of electron density close to the iron center of the active site, which was found to be consistent with the presence of a monodentate O-coordinated CO_2 molecule [13b].

7.1.3.2 Carbon Dioxide as C-Electrophile

η^1-C coordination of CO_2 to metal centers has been clearly demonstrated [12]. The structural data of the adducts indicate an important pouring of electron density from metal-filled d-orbitals into the empty antibonding $2\pi_u$(6a1) orbital, which weakens the C–O bonds and, as expected from the Walsh diagram (Fig. 7.4), causes a remarkable bending of the coordinated CO_2 molecule. Analogous structural changes always mark the geometry of CO_2 molecule whenever the electrophilic carbon atom of CO_2 is involved in an interaction with electron-rich species (hydride, amines, alcohols, carbanions, etc.) [12].

This interaction also affects the nucleophilicity of the O-atoms of CO_2. For instance, in CO_2 adducts with amines or alcohols, the resulting carbamate or carbonate anions, under suitable conditions, can easily react with C-electrophiles to give organic carbamates [14] or carbonates, [15] respectively (see Chaps. 9, 10).

7.1.3.3 Amphoteric Reactivity of Carbon Dioxide

Most frequently, however, the reactivity of CO_2 fully reveals the intrinsically amphoteric nature of the molecule. Side-on (η^2-C, O) coordination of CO_2 to metal

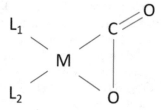

Fig. 7.6 Side-on coordination of CO_2 to a metal center

complexes (Fig. 7.6) exemplifies in a clear way this behavior. Different schemes can be proposed for describing the metal-CO_2 ligand bond in these systems. These adducts may be considered as three-membered oxametallacycles resulting from a formal oxidative addition of one of the π-bonds of the cumulene to the metal center; in this case, both the metal centers, which increase by two units their formal oxidation state, and CO_2 act simultaneously as electron acceptor and electron donor. Otherwise, (η^2-C, O)-adducts can be regarded as π-complexes reminiscent of the metal–olefin bond and involves both electron donation from the O-centered $1\pi_{uy}(5a_1)$ and $1\pi_{gy}(4b_2)$ orbitals of CO_2 to empty d-orbitals of the metal center and, at a greater extent, electron back-donation from a filled d-orbital of the metal to the empty carbon-centered $2\pi_{uy}(6a_1)$ orbital of CO_2 [16]. Whatever is the model, in these metal systems, CO_2 behaves as electron donor through an end-oxygen and electron acceptor at the central carbon [12].

Here, it may be worth mentioning that solid CO_2 [17], which exists in different phases under a pressure of 40 GPa and laser heating (ca. 1800 K), undergoes a dramatic change and the π molecular bonds localized on C=O bonds change into an extended network of C–O single bonds, a new crystalline "super-hard" phase, called CO_2-V [17c], which can be quenched at ambient temperature.

The crystal structure of CO_2-V was found to be orthorhombic ($P2_12_12_1$), analogous to that of SiO_2 tridymite (a distorted high-temperature phase of ß-quartz) [17d]. The new phase is composed of CO_4 tetrahedra, where each carbon atom is bonded to four oxygen atoms at a carbon–oxygen distance of 1.36 Å at 40 GPa and an O–C–O angle of 110°. The C–O–C angle of 130° is markedly smaller than those of SiO_2 tridymites (174–180°) or quartz (145°). CO_2-V is a "super-hard" polymeric form of CO_2.

More recently, it has been reported also the synthesis of an amorphous, silica-like form of carbon dioxide (α-CO_2), which is called "α-carbonia" [17e]. This material is homologous to other Group 14 dioxide glasses (α-silica; α-germania). Both CO_2-V and α-CO_2 are converted back to the molecular state when pressure and temperature are brought back to ambient values.

Recent results suggest that carbon dioxide polymerization does not occur via intermediate states where molecules gradually distort as pressure increases, but is most likely due to solid-state chemical reactions between CO_2 molecules [17g].

7.1.3.4 The Carbon Dioxide Radical Anion, CO_2^-

CO_2^- radical anion represents the first step leading to the reduction of CO_2 (Chap. 9). It is isoelectronic with NO_2 and should have the same geometry. Qualitative Walsh rules [5] predict a bent equilibrium geometry (C_{2v} symmetry) with elongated CO bonds for CO_2^- in the $^2A_1$ ground state, which may be expressed as $(1a_1)^2$ $(1b_2)^2$ $(2a_1)^2$ $(3a_1)^2$ $(2b_2)^2$ $(4a_1)^2$ $(3b_2)^2$ $(1b_1)^2$ $(5a_1)^2$ $(1a_2)^2$ $(4b_2)^2$ $(6a_1)^1$ (see Fig. 7.4). Such view is supported by quantum-mechanical studies [18]. According to a recent theoretical analysis at B3LYP6-31-G**//B3LYP/6-31G* level [18f], the charge density distribution of CO_2^-, from a natural population analysis (NPA), shows negative character for the two O-atoms ($-0.76e$) and a positive charge on the C-atom ($+0.52e$). Spin (densities) populations indicate a 68% electron delocalization on the C-atom and 16% on each oxygen. CO_2^- is, therefore, a radical-like species at carbon and basic at oxygen. Accordingly, calculations show that O-protonation to afford the hydroxycarbonyl radical OCOH˙ is favored over C-protonation, which should generate the less stable formiloxy radical HCOO˙ [18a, g]. These studies also show that *trans*-OCOH˙ is more stable than the *cis*-isomer [18a, g, h].

In the gas phase, CO_2^- is metastable against electron autodetachment (Eq. 7.2).

$$CO_2^- \rightarrow CO_2 + e^- \tag{7.2}$$

The existence of metastable autodetaching CO_2^- anion in the gas phase was first noted by Paulson [19a] by studying the reactions of O^- ion with CO_2. CO_2^- has been generated [19b, c] in the collision of electrons or Cs atoms with cyclic anhydrides (maleic anhydride, succinic anhydride), which contained the bent (ca. 133°) OCO unit. A lifetime of 30–60 μs was measured with a time-of-flight spectrometer. CO_2^- ions formed in collisions of alkali-metal atoms with linear CO_2 molecules have a mean lifetime of 90 ± 20 μs and a value of -0.6 ± 0.2 eV for the adiabatic electron affinity (EA_{ad}) of the CO_2 ground state [19d]. Such EA_{ad} agrees well with the computed value of -0.67 eV and is much lower than the vertical electron affinity (-3.6 eV, determined by electron scattering measurements) [19e]. The large difference between the two values can be justified on the basis of the stabilization of CO_2 upon bending. The relatively high stability of the anion is justified taking into account the barrier (0.4 eV) raised by the change in molecular geometry going from the $^2A_1$ state of CO_2^- to the ground state of CO_2 [18b]. Therefore, the CO_2^- molecule in its equilibrium geometry is metastable, being kinetically stabilized.

CO_2^- in a solid matrix [20a] was generated by irradiation of sodium formate crystals with γ-rays at ambient temperature. ESR experiments allowed to ascertain that OCO angle in CO_2^- is 134° and has a $^2A_1$ ground state with the electron being in the half-filled $6a_1$ molecular orbital (14% carbon 2s, 66% carbon $2p_z$, and 10% for each oxygen $2p_z$). At room temperature, the decay half-life of the radical anion in a KBr disk was estimated to be over 1 year.

CO_2^- easily reacts with water, affording hydrogen carbonate and formate anions (Eq. 7.3).

$$2\ CO_2^- + H_2O \rightarrow HCO_3^- + HCO_2^- \tag{7.3}$$

Under anhydrous conditions, it decays by a bimolecular process to afford CO_3^{2-} and CO [20]. Interestingly, in a KCl matrix, oxalate dianion is observed as a minor product.

The second-order kinetics for the decay of the CO_2^- radical, under anhydrous conditions, is consistent with the mechanism summarized in Scheme 7.1. CO_2^- radicals have been detected by ESR in biological apatites (tooth enamel, bone) and their synthetic analogs exposed to γ-rays or UV light, or subjected to thermal treatment [20d].

The putative mechanism of formation of CO_2^- in such cases is represented in Scheme 7.2. Electrons generated from impurities under radiation are captured by CO_3^{2-} which is transformed into a metastable short-lived CO_3^{3-} radical, which decays to CO_2^- and the oxide dianion.

Single electron reduction of CO_2 to CO_2^- (Eq. 7.4) has a particular interest as it is the first rate determining step in multi-electron electrochemical reduction of CO_2 to other valuable species, such as formic acid, alcohols, hydrocarbons, CO, and oxalate [21].

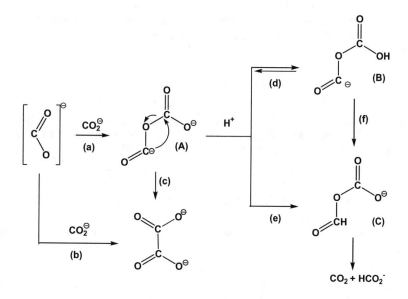

Scheme 7.1 Mechanism of formation of oxalates and $CO + CO_3^{2-}$ from CO_3^{2-} [19b]

$$CO_3^{2-} + e^- \quad \rightarrow \quad CO_3^{3-} \quad \rightarrow \quad CO_2^- + O^{2-}$$

Scheme 7.2 Putative mechanism of formation of CO_3^{2-} in apatites

$$CO_2 + e^- \rightarrow CO_2^- \qquad (7.4)$$

Reaction 7.4 occurs at high voltage $V_{(CO_2/CO_2^-)} = -1.90$ to -2.2 V, depending on whether the reaction occurs in water or organic solvents, respectively. Such high voltage is due to the large reorganizational energy necessary in going from the linear molecule to the bent radical anion (-0.4 eV, see Sect. 7.1.3.4) [22]. Accordingly, CO_2 electrochemical reduction does not occur easily and the really applied electrolysis potentials for CO_2 reduction are more negative than the thermodynamic values. As seen above, CO_2^- has a life depending on several parameters such as concentration of reactants, electrode potential, temperature, electrocatalyst material, and nature of electrolyte solution (i.e., aqueous *versus* non-aqueous electrolyte) that affect, thus, the nature and selectivity of products.

In general, radiations can generate CO_2^- from formate/formic [19a] together with the hydroxycarbonyl radical $OCOH^-$ which pK_a is 2.3 [19b]. The mechanism of decay of CO_2^- is dependent on pH and is shown in Scheme 7.1 [19b]. A major feature of this mechanism is that CO_2^- radicals react mainly (>90%) by head-to-tail recombination to give intermediate **A**, which may rearrange to oxalate or undergo a competing proton-catalyzed disproportionation, which accounts for the formation of CO_2. According to the proposed mechanism, protonation of the intermediate **A** at the oxygen atom (step (d)) should be faster than step (e). However, once **B** is formed, it can undergo subsequent protonation at carbon (step (f)), possibly assisted by a molecule of water, to give the mixed anhydride **C**, which decomposes to CO_2 and formate.

7.1.3.5   The Carbon Dioxide Radical Cation, CO_2^+

Removal of one of the $1\pi_g$ electrons from the CO_2 molecule generates the CO_2^+ radical cation in the ground state $^2\Pi_g$. The energy required for this process is 13.79 eV [4], which is by far larger than that required for the formation of the radical anion CO_2^- (vide infra) or dissociation of neutral CO_2 to CO and atomic oxygen (vide infra). Spectroscopic measurements [23] show that CO_2^+ in its $^2\Pi_g$ ground state is linear with $D_{\infty h}$ point group symmetry: the two C–O bonds are equivalent with a length which is only slightly longer (1.1769 Å) than that measured in the neutral molecule.

If one electron is removed from the inner orbitals $1\pi_u$, $3\sigma_u$, $4\sigma_g$ of CO_2, a higher amount of energy [4] is required that produces the CO_2^+ radical cation in the excited states $^2\Pi_u$, $^2\Sigma_u^+$, and $^2\Sigma_g^+$, respectively, all linear ($D_{\infty h}$) [23, 24]. The experimental C–O bond distances for the $^2\Pi_u$ and $^2\Sigma_u^+$ excited states are, respectively, 1.228 Å and 1.180 Å. The available structural data, as well as theoretical calculations, [24] show that a significant increase in the carbon–oxygen bond length takes place, upon ionization, when the electron is removed from the CO_2 $1\pi_u$ orbital, which is strongly bonding. Conversely, a modest change is observed when the electron is removed from the orbitals $1\pi_g$, $3\sigma_u$, or $4\sigma_g$, which exhibit, mainly or fully, non-bonding character.

Carbon dioxide cation plays an important role in the dynamics of plasma discharges and chemistry of planetary atmospheres [25]. Multiple ionization of the molecule (CO_2^{n+}) might also occur in collisional processes that might be followed by Coulomb explosion (Eq. 7.5), that is, the fragmentation of the molecular ion into two or three atomic ions [26].

$$CO_2^{n+} \rightarrow C^{p+} + O^{q+} + O^{r+} \tag{7.5}$$

7.2 Spectroscopic Techniques Applied to the CO_2 States

7.2.1 Infrared (and Raman) Spectroscopy

The vibrational (and rotational) spectra of carbon dioxide have been extensively investigated for

- their relevance to the study of planetary atmosphere [27];
- recording of CO_2 level in the air and analysis of remote sensing data; and
- detecting the bonding state of the COO moiety in chemicals and materials [28, 29].

The linear triatomic molecule in its ground state exhibits four normal vibration modes (Fig. 7.7) according to the (3N–5) rule, where N is the number of atoms in the molecule. While the antisymmetric stretching (v_3) and the doubly degenerate bending (v_2) modes are IR-active, the symmetric stretching mode (v_1) is IR-inactive (but is active in Raman spectroscopy), as this normal mode does not generate any change of the electric dipole-moment of the molecule. In the IR spectrum of free (gaseous) CO_2, the antisymmetric stretching frequency (v_3) is found at 2349.16 cm^{-1}, [30a] while it is found at 2344.0 cm^{-1} in the solid state (15 K) [30b]. In aqueous solution, this stretching (2343 cm^{-1}) is shifted by only 6 cm^{-1} from the vapor phase, indicating the absence of hydrogen bonding between water molecules and dissolved CO_2 [30c]. The infrared absorption attributed to the bending vibration of the molecule is observed at 667.38 cm^{-1} for gaseous CO_2 [30] and at 654.7 and 659.8 cm^{-1} for solid CO_2, with a full width at half maximum of

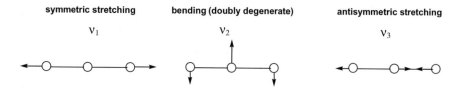

symmetric stretching   **bending (doubly degenerate)**   **antisymmetric stretching**

v_1 v_2 v_3

Fig. 7.7 Normal vibration modes of the carbon dioxide molecule in its ground state. Reprinted from Ref. [5b], Copyright (2016), with permission from Elsevier

1.8 and 3.1 cm^{-1}, respectively, showing Davydov splitting due to a high-ordered dry-ice structure [30b]. Because carbon dioxide molecule has a center of symmetry, the antisymmetric stretching (v_3) and the bending (v_2) modes are Raman inactive. A Fermi dyad is observed in the Raman spectrum of gaseous CO_2 at 1285.40 and 1388.15 cm^{-1} [30a], which originates from the resonance between the unperturbed energy levels associated with the fundamental transition of the v_1 totally symmetric stretching mode ($v_1 = 1333$ cm^{-1}) and the harmonic transition $2v_2$ of the v_2 bending mode.

Table 7.1 compares the IR fundamental frequencies of gaseous CO_2 [30a] with those measured for gaseous CO_2^+ [31] and for CO_2^- isolated in solid neon [32].

The IR spectrum of CO_2^- in a neon matrix, where the radical anion is free from interactions with metal cations, exhibits marked differences with respect to that of the neutral parent molecule. Due to the bent geometry (C_{2v}) of the anion, all the three normal vibration modes of CO_2^- are IR-active. The changes of the stretching frequencies are pronounced. Their lower values reflect a reduced CO bond order, which is 1.5 in CO_2^- as compared with 2 for CO_2. It is worth noting that the interaction of the radical anion with a metal cation can cause significant shifts of the frequency values v_1, v_2, and v_3 with respect to those tabulated in Table 7.1 [20b, 33].

The value observed for the antisymmetric stretching frequency v_3 of gaseous CO_2^+ is very close to that measured for the radical cation in solid neon ($v_3 = 1421.7$ cm^{-1}) [32], an anomalously low value [34]. Such anomaly is explained [34a] in terms of a vibronic interaction between the $^2\Pi_g$ ground state and the $^2\Pi_u$ excited electronic state through the v_3 vibration normal mode.

It is worth of note that the formation of CO_2-E (E = heteroatom) adducts may involve the population of one of the LUMOs of CO_2 that can originate large modifications in the IR spectrum of the CO_2 moiety. The antisymmetric stretching mode, $v_a(OCO)$, is lowered in the range 2250–1400 cm^{-1}, the symmetric stretching mode, $v_s(OCO)$, becomes IR-active and can absorb in the region 1400–1100 cm^{-1}, the bending mode $\delta(OCO)$ is shifted from 667 cm^{-1}, and additional vibrational modes, such as E-carbon and/or E-oxygen stretching modes, C=O out-of-plane deformation, may be observed in the low-frequency region (down to 300 cm^{-1}).

Table 7.1 Vibrational normal modes and related frequencies (cm^{-1}) for neutral, cationic, and anionic CO_2

	v_1 ($v_s(OCO)$)	v_2 ($\delta(OCO)$)	v_3 ($v_a(OCO)$)	Notes	Refs.
CO_2	1333	667.38	2349.16	Gas state	[30a]
CO_2^+	1244.3	511.4	1423.08	Gas state	[31]
CO_2^-	1253.8	714.2	1658.3	Ne matrix	[32]

7.2.2 UV Spectrum of Carbon Dioxide

The UV absorption spectrum of carbon dioxide has been widely studied both experimentally [35] and theoretically [36] for identifying the excited states involved in the electronic transitions. To this end, electron impact spectroscopy has proved to be an additional useful diagnostic tool [37]. Due to the fact that it is transparent in the visible and near-to-middle ultraviolet regions, at least down to 210 nm ($hv <$ 5.91 eV), both in the liquid phase [38] and in the gas phase [23a], the UV spectrum has scarce application in synthetic chemistry as diagnostic tool. Only weak absorptions can be observed below 11 eV. CO$_2$ shows three maxima in the vacuum ultraviolet region at 147.5 (8.41 eV), 133.2 (9.31 eV), and 112.1 nm (11.08 eV), respectively. The bands corresponding to the maxima at 147.5 and 133.2 nm exhibit a vibrational structure, which is poor, irregular, and apparently complex for the maximum at 147.5 nm, while is sharper and more regular for that at 133.2 nm. The assignment of these absorption maxima has been controversial for long time. UV spectroscopy is more useful for understanding the photochemistry of CO$_2$ and its molecular dissociation into CO and [O] at 157 nm (Eq. 7.6).

$$CO_2 \rightarrow CO + O \qquad (7.6)$$

Notably, the dissociation of CO$_2$ to CO($^1\Sigma$) + O(^{3}P) violates the spin conservation rule, as the spin of CO$_2$ in the ground state ($^1\Sigma_g^+$) is zero, whereas the total spin of CO($^1\Sigma$) + O(^{3}P) is one and, therefore, such dissociation is a spin-forbidden process. Attempts have been done to explain such situation and the photolysis of CO$_2$ has been carried out under different conditions, using an Hg lamp. The dissociation has been explained on the basis of an electronic transition of the CO$_2$ molecule from the singlet ground state ($^1\Sigma_g^+$) to the upper (bound) singlet state ^{1}B$_2$, which is below the asymptote of the state ^{1}B$_2$, but above the crossover zone with the state ^{3}B$_2$. Dissociation can occur if the molecule can reach the crossover region (by collisions, for instance) and undergo a transition to the triplet state ^{3}B$_2$.

Experiments carried out with labeled C- and O-atoms (^{13}C and ^{17}O) have shown an isotopic enrichment observed in CO and O$_2$ products. Therefore, it has been proposed that the nuclear spin of ^{13}C ($I = 1/2$) and ^{17}O ($I = 5/2$) may introduce additional coupling by hyperfine interaction, which may increase the dissociation rate of the iso-topologues containing ^{13}C and ^{17}O. Interestingly, the ^{17}O enrichment is 2.2 $\pm$ 0.2 times higher than that of ^{13}C, a value very close to the value of 2.7, which gives the $\mu(^{13}$C)/$\mu(^{17}$O) magnetic moment ratio for the two nuclei.

The influence of nuclear spin in photodissociation of molecules is a novel effect which may have important implications and potential applications in the study of terrestrial and planetary atmospheres [39].

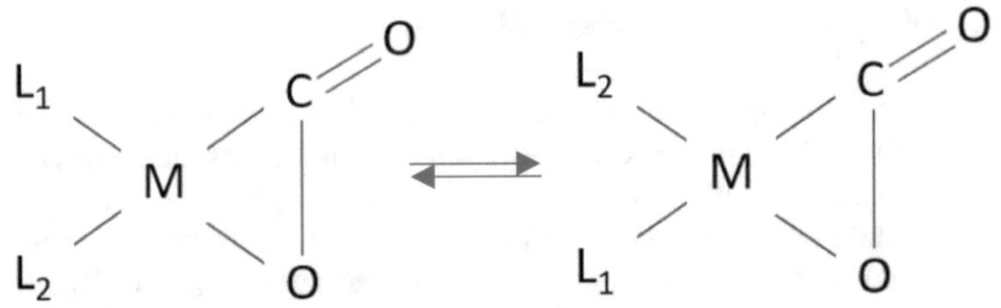

Fig. 7.8 Fluxionality of CO_2 in $Ni(PCy_3)_2(CO_2)$ (M = Ni, L1 = L2 = PCy_3)

7.2.3   Nuclear Magnetic Resonance (NMR) Spectroscopy

CO_2 dissolved in a nonpolar solvent, such as benzene or toluene, exhibits only one ^{13}C resonance around 124 ppm (with respect to tetramethylsilane (TMS)), which moves to 125 ppm in aqueous solutions [40a]. Such resonance is strong enough to be used for the quantification of free CO_2. A water solution distinctly shows the resonance of the hydrogencarbonate anion (HCO_3^-, around 160 ppm) and of the carbonate (CO_3^{2-}, around 168 ppm) [40b].

   ^{13}C-NMR has proved to be a very useful tool for identifying the COO moiety incorporated into a variety of compounds. The ^{13}C resonance of CO_2 shifts down field upon fixation of the heterocumulene into organic products, such as carbamic acids or carbamates (150–160 ppm), organic carbonates (145–165 ppm), organic acids (160–180 ppm) or esters (160–170 ppm), or their metallo-organic analogs (metallo-carbamates $L_nM(O_2CNR_2)$, metallo-carbonates $L_nM(O_2COR)$, metallo-carboxylates $L_nM(O_2CR)$, metallo-carboxylic acids $L_nM(CO_2H)$, metallo-esters $L_nM(CO_2R)$. ^{13}C-NMR has been used with much benefit for the clarification of the structure in solution of the metal-CO_2 adducts and for demonstrating the fluxional behavior of CO_2 adducts in solution. Variable temperature studies have shown that in solution only at low temperature (below 218 K) $Ni(PCy_3)_2(CO_2)$ has a rigid structure that repeats the solid state (see Fig. 7.6) and above 218 K, CO_2 is fluxional with continuous interchange of the ligands (Fig. 7.8) [12].

7.3   The Phase Diagram of CO_2

CO_2 can exist as a gas, liquid, and solid depending on the temperature and pressure. Therefore, it is important to know which phase CO_2 exists under the working conditions. This is shown in Fig. 7.9. CO_2 at ambient conditions (25 °C and 0.1 MPa) is a gas. Its phase diagram (Fig. 7.9) tells how it can exist under conditions different from ambient conditions. As an example, let us consider solid CO_2, commonly known as "*dry ice*," that can be found in ice-cream packages for storing or transport. The phase diagram says that it can exist at −78 °C under ambient pressure (0.1 MPa, blue arrow in the diagram). If one takes a piece of dry ice and

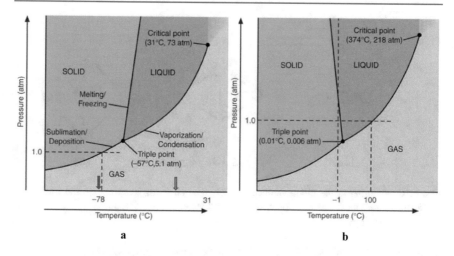

Fig. 7.9   a Phase diagram of CO_2. A point in any colored part of the diagram shows the existence of a single phase (solid, liquid, and gas); a point on a curve represents the equilibrium of two phases (as indicated on the two sides of the curve); the triple point represents the equilibrium of the three phases (solid, liquid, and gas): this is a unique point; the critical point gives the limit T and P of existence of the gaseous phase and the appearance of the supercritical CO_2: a compressed gas that has the properties (density) of a liquid: above such point on the right up square CO_2 cannot be compressed to the liquid phase. **b** Phase diagram of H_2O: note the different shapes and line slopes with respect to the CO_2 diagram

leaves it at ambient conditions (red arrow on the X-axis in the diagram) what will happen? The phase diagram says that solid CO_2 cannot exist under such conditions. Therefore, it spontaneously gets converted into gas (sublimation) as the gaseous phase is the stable phase under ambient conditions. (This is an experiment that everyone can carry out, taking care in handling dry ice that can cause serious damages as it has a temperature of −78 °C and if in touch with skin can quickly dehydrate it causing effects similar to burning: use a pincer for keeping the piece of dry ice. Even, do not place dry ice on temperature-sensitive materials: place it on a piece of cardboard or glass, insulating it from plastic and wood surfaces as these two latter can become fragile and break down. Note that during sublimation, a liquid may appear on the surface where dry ice is placed: it is not liquid CO_2, but condensed water from the atmosphere.)

If one compares water and CO_2, it can be realized that they have quite different properties.

Triple point. H_2O: $T = 0.01$ °C and $P = 0.611$ kPa or 4.58 mmHg; CO_2: $T = -56.8$ °C and $P = 0.51$ MPa

Critical point. Water: $T = 374$ °C and $P = 22.064$ MPa; CO_2: $T = 31.1$ °C and $P = 7.38$ MPa.

The critical point temperature and pressure say that above 31.1 °C whatever pressure is applied gaseous CO_2 will not be converted into liquid, it will remain in a gas-like phase which will be very dense because of the high pressure. Such fluid is

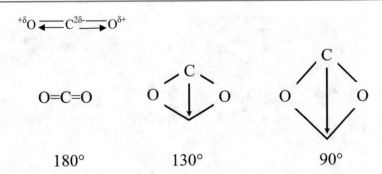

Fig. 7.10 Variation of the dipole moment of CO_2 with changing its geometry: the smaller the OCO angle, more intense is the dipole, indicated by the arrow pointing down

said "*supercritical* CO_2, SC-CO_2" (above the critical parameters) that has interesting properties as solvent and, differently from the gaseous phase, can dissolve liquids and solids affording a single-phase system (gases can dissolve only gases). SC-CO_2 is a low-polarity fluid similar for certain aspects to hexane. It is worth to recall that CO_2 has a linear geometry and is apolar because the two C–O dipoles have equal intensity and point in opposite directions, when CO_2 is pressurized, the geometry can slightly change and the bent molecules will have a non-zero dipole moment (Fig. 7.10). The more the OCO angle deviates from 180°, higher will be the dipole moment.

This is useful because changing the pressure it is possible to modify the polarity of SC-CO_2 and, thus, its solvent properties. Moreover, polar molecules such as alcohols can be dissolved within certain solubility limits in SC-CO_2 so that not only apolar species, but also polar compounds can be dissolved in the SC-phase. This is particularly useful in the extraction of particular chemicals (fragrances or pharmaceuticals or cosmetics) from a matrix like biomass (algae, citrus fruits, tomato peels, etc.) (see also Chap. 8).

All such information is useful for understanding the behavior of CO_2 under various conditions of application.

References

1. (a) Structure data of free polyatomic molecules, Kuchitsu K (Ed) (1992) Landolt-Börnstein, vol. II/21. Springer, Berlin, p. 151; (b) Structure data of free polyatomic molecules, Kuchitsu K (ed) Landolt-Börnstein, vol. II/23. Springer, Berlin, p. 146
2. (a) Vučelić M, Ohrn Y, Sabin (1973) *Ab Initio* calculation of the vibrational and electronic properties of carbon dioxide. J Chem Phys 59:3003–3007; (b) Gutsev GL, Bartlett RJ, Compton RN (1998) Electron affinities of CO_2, OCS, and CS_2. J Chem Phys 108:6756–6762
3. (a) Maroulis G, Thakkar AJ (1990) Polarizabilities and hyperpolarizabilities of carbon dioxide. J Chem Phys 93:4164–4171; (b) Lobue JM, Rice JK, Novick SE (1984) Qualitative structure of $(CO_2)_2$ and $(OCS)_2$. Chem Phys Lett 112:376–380; (c) Johnson MA, Alexander ML, Lineberger WC (1984) Photodestruction cross sections for mass-selected ion clusters:

$(CO_2)_n^+$. Chem Phys Lett 112:285–290; (d) Brigot N, Odiot S, Walmsley SH, Whitten JL (1977) The structure of the carbon dioxide dimer. Chem Phys Lett 49:157–159; (e) Bowen KH, Liesegang GW, Sanders RA, Herschbach DR (1983) Electron attachment to molecular clusters by collisional charge transfer. J Phys Chem 87:557–565; (f) Rossi AR, Jordan KD (1979) Comment on the structure and stability of $(CO_2)_2^-$. J Chem Phys 70:4442–4444; (g) Buckingham AD, Disch RL, Dunmur DA (1968) Quadrupole moments of some simple molecules. J Am Chem Soc 90:3104–3107

4. (a) Allian CJ, Gelius U, Allison DA, Johansson G, Siegbahn H, Siegbahn K (1972) ESCA studies of CO_2, CS_2 and COS. J Elect Spectrosc Relat Phenom 1:131–151; (b) Turner DW (1968) Molecular photoelectron spectroscopy. In: Hill HAO, Day P (eds) Physical methods in advanced inorganic chemistry. Interscience Publishers, London; (c) Turner DW, May DP (1967) Frank-Condon factors in ionization: Experimental measurements using molecular photoelectron spectroscopy II. J Chem Phys 46:1156–1160; (d) Paparo A, Okuda J (2017) Carbon dioxide complexes: Bonding modes and synthetic methods. Coordination Chem Rev 334:136–149

5. (a) Walsh AD (1953) The electronic orbitals, shapes, and spectra of polyatomic molecules. Part II. Non-hydride AB_2 and BAC molecules. J Chem Soc 2266–2288; (b) Aresta M, Dibenedetto A, Quaranta E (2016) State of the art and perspectives in catalytic processes for CO_2 conversion into chemicals and fuels: the distinctive contribution of chemical catalysis and biotechnology. J Cataly 343:2–45

6. (a) Scarlett M, Taylor PR (1986) Protonation of CO_2, COS, CS_2. Proton affinities and the structure of protonated species. Chem Phys 101:17–26; (b) Lias SG, Liebman JF, Levin RD (1984) Evaluated gas phase basicities and proton affinities of molecules. J Phys Chem 13:695–808; (c) Gronert S, Keeffe JR (2007) The protonation of allene and some heteroallenes, a computational study. J Org Chem 72:6343–6352

7. Fock W, McAllister T (1982) Probable abundance ratios for interstellar HCS_2^+, HCOS, HCO_2^+. Astrophys J 257:L99–L101

8. (a) Bogey M, Demuynck C, Destombes JL (1986) The submillimeter wave spectrum of the protonated and deuterated carbon dioxide. J Chem Phys 84:10–15; (b) Bogey M, Demuynck C, Destombes JL (1988) Molecular structure of $HOCO^+$. J Mol Struct 190:465–474

9. (a) Amano T, Tanaka K (1985) Difference frequency laser spectroscopy of the v_1 band of $HOCO^+$. J Chem Phys 82:1045–1046; (b) Amano T, Tanaka K (1985) Difference frequency laser spectroscopy of the v_1 fundamental band of $HOCO^+$. J Chem Phys 83:3721–3728

10. Taddeus P, Guélin M, Linke RA (1981) Three new "nonterrestrial molecules". Astrophys J 246:L41–L45

11. Hammami K, Jaidane N, Lakhdar ZB, Spielfeldel A, Feautrier N (2004) New *ab initio* potential energy surface for the ($HOCO^+$-He) van der Waals complex. J Chem Phys 121:1325–1330

12. Aresta M, Dibenedetto A, Quaranta E (2016) Reaction mechanisms in carbon dioxide conversion. Springer

13. (a) Mauser H, King WA, Gready JE, Andrews TJ (2001) CO_2 fixation by Rubisco: computational dissection of the key steps of carboxylation, hydration, and C–C bond cleavage. J Am Chem Soc 123:10821–10829; (b) Lee HJ, Lloyd MD, Harlos K, Clifton IJ, Baldwin JE, Schofield CJ (2001) Kinetic and crystallographic studies on deacetoxycephalosporin C synthase (DAOCS). J Mol Biol 308:937–948

14. (a) Aresta M, Quaranta E (1997) Carbon dioxide: a substitute for phosgene. ChemTech 27:32–40; (b) Quaranta E, Aresta M (2010) The chemistry of N-CO_2 bonds: synthesis of carbamic acids and their derivatives, isocyanates, and ureas. In: Aresta M (ed) Carbon dioxide as chemical feedstock. Wiley-VCH Verlag GmbH & Co. KGaA, Weinheim

15. Ballivet-Tkatchenko D, Dibenedetto A (2010) Synthesis of linear and cyclic carbonates. In: Aresta M (ed) Carbon dioxide as chemical feedstock. Wiley-VCH Verlag GmbH & Co. KGaA, Weinheim

16. Sakaki S (1990) Transition-metal complexes of nitrogen, carbon dioxide, and similar small molecules. *Ab-initio* MO studies of their stereochemistry and coordinate bonding nature. Stereochem Organometallic Inorganic Compounds (Stereochem. Control., Bonding Steric Rearrange) 4:95–177

17. (a) Santoro M (2010) Non-molecular carbon dioxide at high pressure. In: Boldyreva E, Dera P (eds) High-pressure crystallography: from fundamental phenomena to technological applications. Springer, Dordrecht, The Nederlands; (b) Schettino V, Bini R, Ceppatelli M, Ciabini L, Citroni M (2005) Chemical reactions at very high pressure. Adv Chem Phys 11:105–242; (c) Iota V, Yoo CS, Cynn H (1999) Quartzlike carbon dioxide: an optically nonlinear extended solid at high pressures and temperatures. Science 283:1510–1513; (d) Yoo CS, Cynn H, Gygi F, Galli G, Iota V, Nicol M, Carlson S, Häusermann D, Mailhiot C (1999) Crystal structure of carbon dioxide at high pressure: "superhard" polymeric carbon dioxide. Phys Rev Lett 83:5527–5530; (e) Santoro M, Gorelli FA, Bini R, Ruocco G, Scandolo S, Crichton WA (2006) Amorphous silica-like carbon dioxide. Nature 441:857–860; (f) Yota V, Yoo CS, Klepeis JH, Jenei Z, Evans W, Cynn H (2007) Six-fold coordinated carbon dioxide VI. Nat Mat 6:34–38; g) Datchi F, Giordano VM, Munsch P, Saitta AM (2009) Structure of carbon dioxide phase IV: breakdown of the intermediate bonding state scenario. Phys Rev Lett 103:185701

18. (a) Matoušek I, Fojtík A, Zahradník R (1975) A semiempirical molecular orbital study of radicals and radical ions derived from carbon oxides. Coll Czech Chem Commun 40:1679–1685; (b) Pacansky J, Wahlgren U, Bagus PS (1975) SCF *ab intio* ground state energy surface for CO_2 and CO_2^-. J Chem Phys 62:2740–2744; (c) England WB, Rosemberg BJ, Fortune PJ, Wahl AC (1976) *Ab initio* vertical spectra and linear bent correlation diagrams for the valence states of CO_2 and its singly charged ions. J Chem Phys 65:684–691; (d) England WB (1981) Accurate *ab initio* SCF energy curves for the lowest electronic states of CO_2/CO_2^-. Chem Phys Lett 78:607–613; (e) Sommerfeld T, Meyer HD, Cederbaum LS (2004) Potential energy surface of CO_2^- anion. Phys Chem Chem Phys 6:42–45; (f) Villamena FA, Locigno EJ, Rockenbauer A, Hadad CM, Zweier JL (2006) Theoretical and experimental studies of the spin trapping of inorganic radicals by 5,5-dimethyl-1-pyrroline *N*-oxide (DMPO) 1. Carbon dioxide radical anion. J Phys Chem 110:13253–13258; (g) Feller D, Dixon DA, Francisco JS (2003) Coupled cluster theory determination of the heats of formation of combustion-related compounds: CO, HCO, CO_2, HCO_2, HOCO, HC(O)OH, and HC(O)OOH. J Phys Chem 107:1604–1617; (h) Dixon DA, Feller D, Francisco JS (2003) Molecular structure, vibrational frequencies, and energetics of the HCO, HOCO and HCO_2 anions. J Phys Chem A 107:186–190

19. (a) Paulson JF (1970) Some negative-ion reactions with CO_2. J Chem Phys 52:963–964; (b) Cooper CD, Compton RN (1972) Metastable anions of CO_2. Chem Phys Lett 14:29–32; (c) Cooper CD, Compton RN (1973) Electron attachment to cyclic anhydrides and related compounds. J Chem Phys 59:3550–3565; (d) Compton RN, Reinhardt PW, Cooper CD (1975) Collisional ionization of Na, K, and Cs by CO_2, COS, and CS_2: molecular electron affinities. J Chem Phys 63:3821–3827; (e) Boness MJW, Schulz GJ (1974) Vibrational excitation in CO_2 *via* the 3.8-eV resonance. Phys Rev A 9:1969–1979

20. (a) Vestad TA, Gustafsson H, Lund A, Hole EO, Sagstuen E (2004) Radiation-induced radicals in lithium formate monohydrate ($LiHCO_2.H_2O$). EPR and ENDOR studies of X-irradiated crystal and polycrystalline samples. Phys Chem Chem Phys 6:3017–3022; (b) Symons MCR, West DX, Wilkinson JG (1976) Radiation damage in thallous formate and acetate: charge transfer from thallous ions. Int J Radiat Phys Chem 8:375–379; (c) Hisatsune IC, Adl T, Beahm EC, Kempf RJ (1970) Matrix isolation and decay kinetics of carbon dioxide and carbonate anion free radicals. J Phys Chem 74:3225–3231; (d) Hartman KO, Hisatsune IC (1967) Kinetics of oxalate ion pyrolysis in a potassium bromide matrix. J Phys Chem 71:392–396; e) Chantry GW, Whiffen DH (1962) Electronic absorption spectra of CO_2^- trapped in γ-irradiated crystalline sodium formate. Mol Phys 5:189–194

21. (a) Barton Cole E, Bocarsly AB (2010) Photochemical, electrochemical, photoelectrochemical reduction of carbon dioxide. In: Aresta M (ed) Carbon dioxide as chemical feedstock. Wiley-VCH Verlag GmbH & Co. KGaA, Weinheim; (b) Li W (2010) Electrocatalytic reduction of CO_2 to small organic molecule fuels on metal catalysts. In: Hu Y (ed) Advances in CO_2 conversion and utilization. ACS Symposium Series, American Chemical Society, Washington, DC, pp. 55–76; (c) Gennaro A, Isse AA, Severin M-G, Vianello E, Bhugun I, Savéant J-M (1996) Mechanism of the electrochemical reduction of carbon dioxide at inert electrodes in media of low proton availability. J Chem Soc Faraday Trans 92:3963–3968

22. Wardman P (1989) Reduction potentials of one-electron couples involving free radicals in aqueous solutions. J Chem Ref Data 18:1637–1756

23. (a) Herzberg G (1966) Molecular spectra and molecular structure. III. Electronic spectra and electronic structure of polyatomic molecules. Van Nostrand-Reinhold, New York; (b) Johnson MA, Rostas J (1995) Vibronic structure of the CO_2^+ ion: reinvestigation of the antisymmetric stretch vibration in the X, $\tilde{A}$, and B states. Mol Phys 85:839–868; (c) Gauyacq D, Larcher C, Rostas J (1979) The emission spectrum of the CO_2^+ ion: rovibronic analysis of the $\tilde{A}^2\Pi_u - X^2\Pi_g$ band system. Can J Phys 57:1634–1649; (d) Gauyacq D, Horani M, Leach S, Rostas J (1975) The emission spectrum of the CO_2^+ ion: $B^2\Sigma_u^+ - X^2\Pi_g$ band system. Can J Phys 53:2040–2059

24. (a) Horsley JA, Fink WH (1969) Study of the electronic structure of the ions CO_2^+ and N_2O^+ by the LCAO-MO-SCF method. J Phys B (Atom Mol Phys) Ser 2(2):1261–1270; (b) Grimm FA, Larsson M (1984) A theoretical investigation on the low lying electronic states of CO_2^+ in both linear and bent configurations. Physica Scripta 29:337–343; (c) Chambaud G, Gabriel W, Rosmus P, Rostas J (1992) Ro-vibronic states in the electronic ground state of CO_2^+ ($X^2\Pi_g$). J Phys Chem 96:3285–3293; (d) Gellene GI (1998) CO_2^+: a difficult molecule for electron correlation. Chem Phys Lett 287:315–319

25. (a) Yang M, Zhang L, Zhuang X, Lai L, Yu S (2008) The [1 + 1] two-photon dissociation spectra of CO_2^+ via $\tilde{A}^2\Pi_{u,1/2}$ (v_1v_20) $X^2\Pi_{g,1/2}$ (000) transitions. J Chem Phys 128:164308, 1–7; (b) King SJ, Price SD (2008) Electron ionization of CO_2. Int J Mass Spectrom 272:154–164; (c) Liu J, Chen W, Hochlaf N, Qian X, Chang C, Ng CY (2003) Unimolecular decay pathways of state-selected CO_2^+ in the internal energy range of 5.2–6.2 eV: an experimental and theoretical study. J Chem Phys 118:149–163

26. Siegmann B, Werner U, Lutz HO, Mann R (2002) Complete coulomb fragmentation of CO_2 in collisions with 5.9 MeV u^{-1} Xe^{18+} and Xe^{43+}. J Phys B (At Mol Opt Phys) 35:3755–3766

27. (a) Rothman LS, Hawkins RL, Wattson RB, Gamache RR (1992) Energy levels, intensities, and linewidths of atmospheric carbon dioxide bands. J Quant Spectrosc Radiat Transfer 48:537–566; (b) Fox K (1972) High resolution infrared spectroscopy of planetary atmospheres. In: Rao KN, Mathews CW (eds) Molecular spectroscopy: modern research. Academic press, New York; (c) White DW, Gerakines PA, Cook AM, Whittet DCB (2009) Laboratory spectra of the CO_2 bending-mode feature in interstellar ice analogues subject to thermal processing. Astrophys J Suppl S 180:182–191

28. Gibson DH (1996) The organometallic chemistry of carbon dioxide. Chem Rev 96:2063–2095

29. (a) Jegat C, Fouassier M, Mascetti J (1991) Carbon dioxide coordination chemistry. 1. Vibrational study of trans-Mo(CO$_2$)$_2$(PMe$_3$)$_4$ and Fe(CO$_2$) (PMe$_3$)$_4$. Inorg Chem 30:1521–1529; (b) Jegat C, Fouassier M, Tranquille M, Mascetti J (1991) Carbon dioxide coordination chemistry. 2. Synthesis and FTIR study of Cp$_2$Ti(CO$_2$) (PMe$_3$). Inorg Chem 30:1529–1536; (c) Jegat C, Fouassier M, Tranquille M, Mascetti J, Tommasi I, Aresta M, Ingold F, Dedieu A (1993) Carbon dioxide coordination chemistry. 3. Vibrational, NMR, and theoretical studies of Ni(CO$_2$)(PCy$_3$)$_2$. Inorg Chem 32:1279–1289

30. (a) Shimanouchi T (1972) Tables of molecular vibrational frequencies, consolidated Volume I. NSRDS-NBS (US) 39:1–164; (b) van Broekhuizen FA, Groot IMN, Fraser HJ, van Dishoeck EF, Schlemmer S (2006) Infrared spectroscopy of solid CO–CO$_2$ mixtures and layers. A&A 451:723–731; (c) Falk M, Miller AG (1992) Infrared spectrum of carbon dioxide in aqueous solution. Vibr Spectrosc 4:105–108

31. Jacox ME (1990) Vibrational and electronic energy levels of polyatomic transient molecules. Supplement 1. J Phys Chem Ref Data 19:1388–1546

32. (a) Jacox ME, Thompson WE (1989) The vibrational spectra of molecular ions in solid neon CO_2^+ and CO_2^-. J Chem Phys 91:1410–1416; (b) Jacox ME, Thompson WE (1999) The vibrational spectra of CO_2^+, $(CO_2)_2^+$, CO_2^- and $(CO_2)_2^-$ trapped in solid neon. J Chem Phys 110:4487–4496; (c) Zhou M, Andrews L (1999) Infrared spectra of the CO_2^- and $C_2O_4^-$ anions in solid argon. J Chem Phys 110:2414–2422

33. (a) Jacox ME, Milligan DE (1974) Vibrational spectrum of CO_2^- in an argon matrix. Chem Phys Lett 28:163–168; (b) Kafafi ZH, Hauge RH, Billups WE, Margrave JL (1983) Carbon dioxide activation by lithium metal. 1. Infrared spectra of $Li^+CO_2^-$, $Li^+C_2O_4^-$ and $Li_2^{2+}CO_2^{2-}$ in inert gas matrices. J Am Chem Soc 105:3886–3893; (c) Manceron L, Loutellier A, Perchard JP (1985) Reduction of carbon dioxide to oxalate by lithium atoms: a matrix isolation study of the intermediate steps. J Mol Struct 129:115–124; (d) Kafafi ZH, Hauge RH, Billups WE, Margrave JL (1984) Carbon dioxide activation by alkali metals. 2. Infrared spectra of $M^+CO_2^-$ and $M_2^{2+}CO_2^{2-}$ in argon and nitrogen matrices. Inorg Chem 23:177–183; (e) Bencivenni L, D'Alesssio L, Raimondo F, Pelino M (1986) Vibrational spectra and structure of $M(CO_2)$ and $M_2(CO_2)_2$ molecules. Inorg Chim Acta 121:161–166

34. (a) Kawaguchi K, Yamada C, Hirota E (1985) Diode laser spectroscopy of the CO_2^+ v_3 band using magnetic field modulation of the discharge plasma. J Chem Phys 82:1174–1177; (b) Carter S, Handy NC, Rosmus P, Chambaud G (1990) A variational method for the calculation of spin-rovibronic levels of Renner-Teller triatomic molecules. Mol Phys 71:605–622; (c) Johnson MA, Rostas J (1995) Vibronic structure of the CO_2^+ ion: reinvestigation of the antisymmetric stretch vibration in the X, Ã, and B states. Mol Phys 85:839–868

35. (a) Rabalais JW, McDonald JM, Scherr V, McGlynn SP (1971) Electron spectroscopy of isoelectronic molecueles. II. Linear triatomic groupings containing sixteen valence electrons. Chem Rev 71:73–108; (b) Spielfieldel A, Feautrier N, Cossart-Magos C, Werner H-J, Botschwina P (1992) Bent valence states of CO_2. J Chem Phys 97:8382–8388; (c) Cossart-Magos C, Launay F, Parkin JE (1992) High resolution absorption spectrum of CO_2 between 1750 and 2000 Å. 1. Rotational analysis of nine perpendicular-type bands assigned to a new bent-linear electronic transition. Mol Chem Phys 75:835–856; (d) Wang Y-G, Wiberg KB, Werstiuk NH (2007) Correlation effects in EOM-CCSD for the excited states: evaluated by AIM localization index (LI) and delocalization index (DI). J Phys Chem 111:3592–3601; (e) Winter NW, Bender CF, Goddard III WA (1973) Theoretical assignments of the low-lying electronic states of carbon dioxide. Chem Phys Lett 20:489–492

36. (a) England WB, Ermler WC (1979) Theoretical studies of atmospheric triatomic molecules. New *ab initio* results for the $^1\Pi_g$–$^1\Delta_u$ vertical state ordering in CO_2. J Chem Phys 70:1711–1719; (b) Spielfeldel A, Feautrier N, Chambaud G, Rosmus P, Werner H-J (1993) The first dipole-allowed electronic transition of $1\,^1\Sigma_u^+ - X\,^1\Sigma_g^+$ of CO_2. Chem Phys Lett 216:162–166; (c) Buenker RJ, Honigmann M, Liebermann H-P, Kimura M (2003) Theoretical study of the electronic structure of carbon dioxide: bending potential curves and generalized oscillator strengths. J Chem Phys 113:1046–1054; (d) Wiberg KB, Wang Y-G, de Oliveira AE, Perera SA, Vaccaro PH (2005) Comparison of CIS and EOM-CCSD-calculated adiabatic excited states structures. Change in charge density on going to adiabatic excited states. J Phys Chem 109:466–477

37. (a) Lasettre EN, Skerbele A, Dillon MA, Ross KJ (1968) High-resolution study of electron-impact spectra at kinetic energies between 33 and 100 eV and scattering angles to 16°. J Chem Phys 48:5066–5097; (b) McDiarmid R, Doering JP (1984) Electronic excited states of CO_2: an electron impact investigation. J Chem Phys 80:648–656; (c) Chan WF, Cooper G, Brion CE (1993) The electronic spectrum of carbon dioxide. Discrete and continuum photoabsorption oscillator strengths (6–203 eV). Chem Phys 178:401–413

38. Eiseman Jr BJ, Harris L (1932) The transmission of liquid carbon dioxide. J Am Chem Soc 54:1782–1784

39. (a) Okabe H (1978) Photochemistry of small molecules. Wiley, Inc., New York; (b) Slanger TG, Black G (1978) CO_2 photolysis revised. J Chem Phys 68:1844–1849; (c) Zhu Y-F, Gordon RJ (199) The production of $O(^3P)$ in the 157 nm photodissociation of CO_2. J Chem Phys 92:2897–2901; (d) Matsumi Y, Shafer N, Tonukura K, Kawasaki M, Huang Y-L, Gordon RJ (1991) Doppler profiles and fine structure branching ratios of $O(^3P_J)$ from photodissociation of carbon dioxide at 157 nm. J Chem Phys 95:7311–7316; (e) Miller RL, Kable SH, Houston PL, Burak I (1992) Product distributions in the 157 nm photodissociation of CO_2. J Chem Phys 96:332–338; (f) Mahata S, Bhattacharya SK (2009) Anomalous enrichment of ^{17}O and ^{13}C in photodissociation products of CO_2: possible role of nuclear spin. J Chem Phys 130:234312, 1–17

40. (a) Liger-Belair G, Prost R, Parmentier M, Jeandet P, Nuzillard J-M (2003) Diffusion coefficient of CO_2 molecules as determined by ^{13}C NMR in various carbonated beverages. J Agric Food Chem 51:7560–7563. (b) Holmes PE, Naaz M, Poling BE (1998) Ion Concentrations in the CO_2, NH_3, H_2O System from ^{13}C NMR Spectroscopy. Indus Eng Chem Res 37:3281–3287

Use of CO$_2$ as Technical Fluid (Technological Uses of CO$_2$)

8

Abstract

This Chapter describes the "non-chemical" uses of CO$_2$, which means all those applications in which CO$_2$ is not converted into other chemicals, but remains the same species. Some of the applications are quite familiar to the public (food packaging, low temperature cooling agent, fluid as fire extinguisher, cleaning fluid, smoke effects, and water treatment), others are typical of industry and less known to the public. In water treatment, indeed, CO$_2$ is converted into its hydrated forms; this is the only case in which CO$_2$ loses its original nature, as it happens anytime CO$_2$ is dissolved in water.

8.1 Introduction

CO$_2$ is commercialized as a *liquid* in cylinders (vapor pressure *ca.* 6.5 MPa), from which either gas (from the top) or liquid (from bottom) can be spilled, or as a *solid* (dry ice). A popular use of liquid and solid CO$_2$ is cooling of a variety of goods and food that can be carried out in closed compartments (batch) or else in continuous apparatuses (tunnels through which goods are passed). In storage compartments, or even in transport vans, L-CO$_2$ is sprayed from a cylinder; L-CO$_2$ is converted into icy-snow (that covers the surface of the compartment or even of the goods). Alternatively, dry ice in blocks is used: both icy-snow and blocks upon sublimation take heat from the environment which is cooled down.

Parts of this chapter were reproduced with permission from Ref. [20]. Copyright 2019, Springer Nature.

© Springer Nature Switzerland AG 2021
M. Aresta and A. Dibenedetto, *The Carbon Dioxide Revolution*,
https://doi.org/10.1007/978-3-030-59061-1_8

The technological uses of CO_2 include all those applications in which CO_2 is used as fluid (gas and liquid) or solid, is not transformed into another chemical, maintains its status of triatomic molecule during its application, and at the end is either vented to the atmosphere or captured and reused. Such uses can be categorized as follows:

- Food Industry (cooling agent, food packaging, antibacterial)
- Additive to beverages
- Extraction of fragrances
- Solvent and processing fluid in the chemical industry
- Dry-cleaning fluid
- Cleaning agent in metal- and electronic-industry
- Fluid in air-conditioners
- Fire extinguisher
- Water treatment (the only case in which CO_2 is converted into HCO_3^- or CO_3^{2-})
- Greenhouses and agriculture
- Use as dry ice
- Other uses.

The whole amount of CO_2 used in such operations (and EOR, already discussed in Chap. 6) is known as the *Merchant Market of CO_2* and amounts at *ca.* 30 $MtCO_2$ as for 2019.

8.2   Sources of CO_2

So far, the majority, if not all, CO_2 used as technological fluid was extracted from natural wells at a very low cost (*ca.* 10–15 US\$/t). Such operation aggravates the CO_2 immission into the atmosphere. Therefore, such natural CO_2 is being substituted these days with CO_2 separated from industrial gas streams and power stations flue gases. Capture, purification, and liquefaction of carbon dioxide is capital and energy intensive. The demand of CO_2 is such that it must be continuously available all year long with seasonal peaks during summer. Often it happens that the market suffers shortage, especially of high-quality product, due to, for example, seasonal stop of sources (maintenance of plants). This causes instability of price. However, capture from industrial emissions is environmentally beneficial, but more costly and can suffer discontinuity, while extraction of CO_2 from natural wells is economically more convenient and guarantees continuity of delivery, but it produces an environmental burden.

For uses categorized above, the major sources are concentrated streams from ammonia production, fermentation processes for the production of alcohol, and the chemical/petrochemical industry (e.g., H_2 production) (Table 8.1) [1]. Power stations are continuous sources, but the concentration (4–15%) of CO_2 and its quality are rather low. Industrial sources (Table 8.1) are more concentrated (up to 90$^+$%) and of much better quality (NOx and SOy are absent).

Industrial sector	Mt$_{CO2}$/y produced
Oil Refineries	850–900
Ethene and other Petrochemical Processes	155–300
LNG Sweetening	25–30
Ethene oxide	10–15
Ammonia	160
Fermentation	>200
Cement	> 1000
Iron and steel	ca. 900

Table 8.1 Industrial sources of CO$_2$, other than power Stations. Reproduced from Ref [1] Copyright 2014, publisher publiSQB

Together, the ammonia industry and the H$_2$ industry supply more than 85% of the carbon dioxide commercialized in the merchant market, the rest being mainly CO$_2$ extracted from natural wells, still used as back option in case of shortage.

As seen in Chap. 2, the Syngas (Eq. 8.1) and WGS (Eq. 8.2) processes coproduce hydrogen and carbon dioxide, which are then separated. Hydrogen is used for a variety of applications, among which ammonia synthesis (Eq. 8.3).

$$CH_4 + H_2O \rightarrow CO + 3H_2 \qquad (8.1)$$

$$CO + 3H_2 + H_2O \rightarrow CO_2 + 4H_2 \qquad (8.2)$$

$$N_2 + 3H_2 \rightarrow 2NH_3 \qquad (8.3)$$

$$2NH_3 + CO_2 \rightarrow H_2NC(O)NH_2 + H_2O \qquad (8.4)$$

CO$_2$ recovered in reaction 8.2 is used mainly for urea manufacture (Eq. 8.4). A small quantity is purified and liquefied for sale on the merchant market as liquid CO$_2$. Noteworthy, the ammonia process as a source may not be continuously available all year long due to plant maintenance. Consequently, dramatic shortages of L-CO$_2$ can be experienced especially during summer, as it happened in 2015 in US and during late spring and the summer of 2018 in Western Europe [2] and Mexico. Similarly, during the last few years, even Japan has experienced severe shortages of L-CO$_2$ and dry ice in the early summer periods. It happens, in fact, that during summer (low season for NH$_3$ synthesis) the maintenance of up to 50% of all ammonia plants takes place. But summer is the season that consumers most require liquid and solid CO$_2$. *The production and consumption are, thus, out of phase.* In order to avoid such outages, countries which have differentiated sources are evaluating their exploitation. For example, CO$_2$ from ethanol production (USA and Brazil, essentially) is an excellent candidate for its high concentration and purity. Even biogas produced from waste fermentation is a putative source; in this case, a careful purification is necessary in order to avoid that pathogens and pollutants are imported into production chains and food. New sources such as the TPI (Siad) plant in the Netherlands are under scrutiny for recovering CO$_2$ from the flue gases emitted by the waste-to-energy plant [3]. Another example is the Quad-Generation approach being promoted especially in regions with limited or unreliable CO$_2$ supply. Among others, Coca Cola has been implementing this in some of its bottling plants [4]. The Direct Air Capture, discussed in Chap. 6 would solve this outage issue as CO_2 would be produced when and where required [5], but such technology at the moment has a cost not economically viable for such an application. Other initiatives such as NewCO$_2$ [6] try to fill the gap between available CO_2 production sites and ad hoc demands.

8.3 Uses of CO_2 and Its Quality

Different applications demand each a specific purity of CO_2. The CO_2 source determines the nature of contaminants, such as nitrogen- or sulphur-oxides in flue gases or hydrocarbons in some of the industrial streams listed in Table 8.1, which need to be removed for most applications. Food-grade CO_2 is the one that requires elimination of contaminants at maximum level, and this increases the cost of production. The standards for purity are internationally defined: ISBT, CGA, EIGA define standards for purity. Such standards tend to guarantee that from food-grade CO_2 all possible contaminants have been removed to one part per billion (ppb) level so that they will not influence taste or odor and will be harmless for the health of consumers. Other non-noxious species, such as water (humidity) and oxygen, also need to be removed to a level acceptable for the specific process in which CO_2 is employed. Quality control and quality assurance measures currently implemented by the producers of L-CO$_2$ for food utilization are very strict. Noteworthy, the detection limit of contaminants has much improved in the last years and allows to go down to ppb threshold.

8.3.1 Food Industry

L-CO$_2$ finds a large use in food industry for a variety of applications that demand food-grade CO_2 that does not contain contaminants potentially harmful to humans.

Most common uses are *additive to beverages, cooling agent during processing and transportation, food packaging (modified atmosphere packaging-MAP), extraction of fragrances, slaughter and stunning of pig, poultry confectioning, supercritical extraction, decontamination of food stuff (antibacterial).* As a matter of fact, CO_2 goes across the almost entire foodstuff series of products for their preparation, shipping, selling, and conservation. In cooling applications, with respect to mechanical cooling, L-CO$_2$ saves space as the compressor is avoided.

Safety has a key role in such applications and procedures for Hazard Analysis of Critical Control Points, or HACCP, which have been established to identify the correct steps in the food processing chain.

In food *packaging* applications, CO_2 now competes with dinitrogen (N_2), the choice being driven by the availability (N_2 is obtained by air-distillation and is, in principle, continuously available) and cost (the cost of DAC, which would make CO_2 always available, is much higher than that of N_2 production). Both gases are non-toxic and leave no residue on the product. In packaging uses, carbon dioxide gas, generated from liquid, is used as an inert medium to prevent flavor loss, deterioration by oxidation, and growth of bacteria. Carbon dioxide is used for packaging coffee and in the processing and transporting of fruits, vegetables, and cereals. Liquid carbon dioxide is used to *refrigerate* food both prior and during shipment. It is sometimes used to pre-cool trailers equipped with conventional refrigeration systems. Trailers can either be filled with dry ice or contain special boxes that will accumulate CO$_2$ snow, injected from a liquid supply tank. As said above, when L-CO$_2$ is directly injected into trailers, some of the liquid carbon dioxide also deposits as layers of snow, directly on the food or on surfaces of the containers. The gradual sublimation during shipment provides additional refrigeration. The modest equipment costs associated with the use of carbon dioxide chilling and its rapid effect are the main factors behind carbon dioxide's popularity in this use. The technology for modified or controlled atmosphere packaging (MAP/CAP) is progressing with the employment of breathable flexible films in combination with gas flushing (carbon dioxide, nitrogen, or blends of the two) targeting the extension of the shelf life of packaged foods. Such progress is quite important as it reduces or cancels the use of preservatives and even reduces the formation of waste, while fully meeting the demand of consumers. The use of CO$_2$ cooling is now extending to other goods such as cooked-baked goods, meat, and poultry.

A list of food products packaged, shipped, stored using L-CO$_2$ is: raw red meat and poultry, sausages, sliced meat, raw fish, smoked fish, cooked fish, prawns, hard cheese (pieces and slices), soft cheese, yogurt, and now even lasagna and other pastas (mass derivatives). The temperature of storage ranges from 0 °C to 6 °C, depending on the nature of the food. CO$_2$ is used neat or in mixture with dinitrogen with respect to which offers better and more flexible solutions.

L-CO$_2$ is also used for helping slicing certain food that is sliced while still frozen. Sometimes, L-CO$_2$ is used to cool liquids by direct injection: CO$_2$ evaporates with time without any modification of the properties of the original liquid.

Another application of CO$_2$ is in slaughterhouses for stunning pigs instead of using electrical stunning, causing less animal stress and injuries. Even poultries or other animals can be treated similarly.

Further, SC-CO$_2$ is used for the pasteurization of food and beverages and preserving them from microorganisms growth. Similarly, CO$_2$ is used as non-toxic fumigant (pesticide or fungicide) in the treatment of cereals when they are shipped or stored in silos. The use is very simple: gaseous CO$_2$ is injected in the silos or container and displaces air by lifting (CO$_2$ is heavier than air). To be effective at least 60% CO$_2$ must be present; therefore, silos do not need to be air tight.

CO$_2$ blocks the microorganisms' growth by desiccation. It is safer than usual fumigants such as CH$_3$Br, and microorganisms cannot develop immunity to it. Additionally, carbon dioxide leaves no residue, and thus does not call for aeration procedures before cereals are sold to the market. Carbon dioxide is registered with the EPA as a non-restricted pesticide.

8.3.2   Additive to Beverages

CO_2 (apolar molecule) is sparingly (0.027 g/L as free CO_2 at ambient conditions) soluble in water (polar molecule) in which undergoes the equilibria discussed in Chap. 6. Therefore, in water, CO_2 produces a slightly acidic solution (pH = 5.8–6) and exists mainly in the form of hydrogencarbonate (see Fig. 6.9).

The amount of CO_2 tends to increase with pressure at constant temperature (Fig. 8.1). Carbonated soft drinks and some types of sparkling wines use large amounts of L-CO_2. The former use benefits from a general shift in some geographical areas in consumer preference toward non-alcoholic beverages (that are low in calories, have a sparkling taste, and are served cold) from traditional beverages, including tea, milk, coffee, beer, and distilled spirits. Soft drink consumption had a strong growth over the last decade but is now expected to be somewhat moderate. Beverages are carbonated under controlled conditions by use of a carbonator or saturator fed with pressurized gas evaporated from L-CO_2. Water cooled to about 5 °C is pumped with CO_2 to the top of the carbonator, then it flows over baffles under pressure where it is saturated and made ready for mixing with additives and bottling. The amount of CO_2 retained by the liquid can be easily controlled and ranges from 2.5% for normal carbonated beverages to 4.5% for highly carbonated beverages such as ginger, tonic water or some bottled drinking waters. Interestingly, some mineral waters from volcanic areas may contain too much CO_2; they are degassed and then loaded with a controlled amount of CO_2 using a carbonator. During such operations part of CO_2 is lost, so that the amount of CO_2 effectively used is higher than that stored into the beverages. CO_2 is also used in the

Fig. 8.1 Amount of CO_2 (all forms) dissolved in water with the temperature under ambient pressure [7]

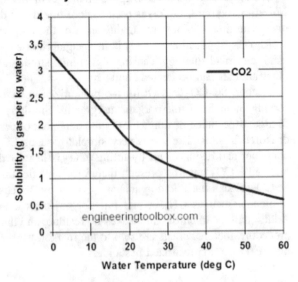

beverage industry as a propellant gas for emptying tanks and as a shielding gas for preserving the drinks quality (exclusion of oxygen).

8.3.3 Extraction of Fragrances

Supercritical-CO$_2$ (SC-CO$_2$) (see Chap. 7) can be used as solvent for the extraction of caffeine from coffee grains or flavors from plants, natural colors, essential oils, active principles from pharmacopea crops. SC-CO$_2$ is contacted with the matrix into a reactor at the wanted temperature (SC-CO$_2$ must be kept at a temperature above 31 °C), then the fluid containing the extracted products (e.g., caffeine) is sent to an expansion reactor where the extracted product (solid or liquid) is released and collected, while CO$_2$ is recovered, recompressed, and recycled. Continuous apparatuses (Fig. 8.2) are available of different volumes for Super Critical Fluids (SCF) extraction. The application of SC-CO$_2$ in decaffeination has met great interest and application because, with respect to the use of organic fluids earlier used (hexane, CH$_2$Cl$_2$), it is not toxic, and does not leave odors and traces.

The extraction of fragrances and lipids is a field of great interest. Bergamot seed oil (*Citrus bergamia* Risso) is currently extracted with SC-CO$_2$, which can be used for lipid extraction from algae. SC-CO$_2$ has been used even for lipid extraction from cheese in order to produce low-fat cheese. Even turmeric extracts from *Curcuma* roots rich in curcuminoids have been obtained by Supercritical Fluid Extraction (SCFE) using a mixture of CO$_2$ and ethanol. From roasted peanuts, the following products have been extracted: hexanal, methylpyrrole, benzene acetaldehyde, methylpyrazine, 2,6-dimethylpyrazine, ethylpyrazine, 2,3-dimethylpyrazine, 2, 3, 5-trimethylpyrazine, 2-ethyl-5-methyl-and 2-ethyl-6-methylpyrazine, 2-furancarboxaldehyde, and 3-ethyl-2,5-dimethylpyrazine.

Fig. 8.2 A continuous SC-CO$_2$ extractor (Authors' Laboratory)

Many other examples can be found in the literature about the use of such supercritical fluid [8].

8.3.4   Solvent and Processing Fluid in the Chemical Industry

SC-CO_2 has been used as solvent in chemical processes. The advantage is that it is possible to modulate the polarity of the solvent by playing on pressure, and the solvent can eventually be easily recovered by expansion, collected, recompressed, and recycled. Such operation demands energy and this is a drawback. In some specific cases, SC-CO_2 can also be used as solvent-reagent. Interestingly, enzymes can be used in SC-CO_2 [9]. Moreover, SC-CO_2 can be even mixed up to a given ratio with solvents such as methanol or ethanol that increases the polarity of the fluid and helps to extract more polar molecules or to build up solvent media of variable polarity [9].

SC-CO_2 has been used also as solvent in polymerization reactions or in modifying polymers.

Liquid and solid carbon dioxide are also used for direct injection into chemical reaction systems to control temperature. As an inert gas, carbon dioxide is used in the chemical industry to purge and fill reaction vessels, storage tanks, and other equipment to prevent the formation of explosive gas mixtures and to protect easily oxidable chemicals from contact with air.

CO_2 is largely used as an alternative to chlorofluorocarbons (CFCs) and hydrocarbons-based blowing agents, which are being phased out due to their allegedly detrimental effect on the Earth's ozone layer, in the 100 kt/y foam industry. Carbon dioxide gas under pressure is introduced into rubber and plastic mixes, and foams the material upon pressure release.

The foam blowing industry is divided into two main markets: thermoplastics and thermosets. Thermoplastics, which are used in residential housing as sheeting and roofing, as well as in commercial roofing applications, are the smaller of the two. Although polystyrene is the biggest section of this market, there are sizable segments in polypropylene and polyvinyl chloride as well as several smaller segments.

Thermosets are a much larger market, taking about four times as much business as thermoplastics. This industry closely follows the plastics market. Polyurethane agents go into residential sheeting, refrigerator doors, insulating foam for pipes, spray foam for electrical outlets, and sandwich panels, and also the thin layer of foam which is placed between layers of metal in walk-in refrigerators, storage trucks, and other larger refrigerating devices. Flashing (i.e., thin protrusions of material at mould joints) can be quickly removed by tumbling these articles with liquid carbon dioxide or crushed dry ice. By injecting liquid CO_2 into the moulded product immediately after the blowing step, the cooling time can be reduced in an important way. Cycle times of existing machines can, therefore, be reduced and the production capacity is increased. Due to increase in performance of today's blowing techniques, the use of CO_2 has reduced in popularity, but still shows significant advantages when used with plastic moulds having an important wall thickness.

Several companies, such as Separex in France and Messer in Germany, have also developed a number of technologies that use supercritical CO$_2$ in the production of extremely fine particles. One recently commercialized process dissolves a variety of compounds ranging from pharmaceuticals to paint pigments in supercritical CO$_2$. When the pressure is suddenly released, the material precipitates as particles smaller than 10 μm with a very homogeneous size distribution.

The same process allows for controlled precipitation of polymers with narrow bands of molecular weight distribution.

8.3.5 Dry-Cleaning Fluid

SC-CO$_2$ can be used as substitute of NH$_3$ or perchloroethene (PERC) in dry cleaning of cloths, a large industrial activity that only in Europe employs 180 000 workers. There is a great environmental benefit in such change as both ammonia and PERC have a larger environmental impact than CO$_2$ itself. The dry-cleaning sector currently has 75 000 cleaning machines using PERC or CFC 113 as cleaning solvents. The substitution with SC-CO$_2$ is not straightforward and requires the development of the right detergents/additives and operative conditions (T, P, fluid-flow, fluid-textile w/w ratio, etc.) in order to guarantee that textiles and garments are not damaged. In particular, it is necessary to guarantee an excellent cleaning effect with soil and stains removal, greying–shrinkage–colorfading avoidance [10].

All such parameters are still under study for a full exploitation of such applications of CO$_2$.

8.3.6 Cleaning Agent in the Metal- and Electronic Industry

SC-CO$_2$ finds diversified applications in the metal and electronic industry. The major use for carbon dioxide in the metal industry is for welding on ships, construction sites, and in manufacturing operations. Carbon dioxide is used alone or with Ar or He as a shielding gas to protect the welding zone from the deleterious effects of oxygen, nitrogen, and hydrogen. Although slightly less effective (He produces the hottest arc and Ar is best cleaner), carbon dioxide finds large use because of its lower cost. Moreover, when used as a shielding gas with semi-automatic micro-wire welding equipment, CO$_2$ speeds up by 10 times the welding speed and cleans the weld surface avoiding wire brushing.

An earlier use (now very limited) was as binding agent, in combination with other additives, in the production of sand moulds and cores for casting iron and other metals.

Messer Group GmbH (Bad Soden, Germany) commercializes a SC-CO$_2$ process for separating sludge formed in metalworking into clean oil (soluble in SC-CO$_2$) and metal particles. Oils, metal, and CO$_2$ can be easily separated, collected, and recycled.

In *spray painting*, CO_2, which mixes well with paint polymers, is used as propellant alternative to VOCs. This technology has been developed by Union Carbide and is marketed under the tradename UNICARB.

Similarly, CO_2 is used since long time as a propellant in many aerosol applications, notably as replacement of CFCs, which are well known to damage the Earth's ozone layer.

SC-CO_2 is used for cleaning electronic components as well as for the removal of hydrocarbon machine coolants from metal parts. Specifically, it is currently being investigated for possible application in a vast range of cleaning, extraction, and thin metal film deposition applications for semiconductors and electronics. It is foreseen to have potential commercial application in the areas of information storage devices, semiconductor devices, electrical and electronic components, precision optical devices, inertial guidance systems, medical equipment and devices, and metal finishing. While competition from established cleaning methods may seem severe, there are a number of attractive features which single out the supercritical carbon dioxide process as a unique method for precision cleaning. These include environmentally compatible cleaning, low carbon dioxide cost and overall cost, rapid processing, ability to clean large areas and complex shapes, ability to form homogeneous cleaning supercritical fluids for general cleans as well as the ability to incorporate additives to tailor more specialized and specific cleaning.

In all cleaning applications, SC-CO_2 guarantees high efficiency and effectiveness in contaminant removal, while responding to the stringent deadlines and aggressive scheduling imposed by legislative bodies to phase out the production and use of ozone-depleting substances such as CFCs and to reduce the noxious Volatile Organic Compounds (VOCs) emission derived from the use of organic solvents. Reducing worker exposure to VOCs is also a mandate in the electronic industry. This is a market that will grow in time.

8.3.7   Fluid in Air-Conditioners

The substitution of CFC in air-conditioners would reduce the environmental impact of air cooling in civil and industrial buildings and in mobile cars. CO_2 has been proposed as an interesting candidate. Norsk Hydro ASA (Oslo) and a Norwegian university have developed a CO_2-based cooling system for automobiles. Daimler Chrysler AG (Stuttgart) has tested the Mobile air Conditioning 2000, and Norsk claims the technology competes favorably in price, weight, space, and energy efficiency with the CFC based one. Likewise, Sanyo Electric Co. (Tokyo) has developed a closed-type rotary compressor that uses CO_2 as coolant agent.

Complete CO_2 based refrigeration systems (YARA-Thermoking) [11] are trying to gain market share at the expense of mechanical refrigeration systems based on CFCs as the former reduce the environmental impact and are less expensive (lower CAPEX and OPEX). As a matter of fact, CO_2-based systems are silent and require little maintenance, reach the target temperature more rapidly and hold it longer.

Unfortunately, the acceptance of this application seems to stagnate with little growth since 2013 due to the lack of re-filling stations for L-CO$_2$.

Linde promotes CO$_2$ as cooling medium for AIRCO systems, e.g., in cars and cooling systems in supermarkets (see, Linde brochure on R744) [12].

In life science blends of 10% Ethylene oxide and 90% CO$_2$ are used for cold sterilization applications [13].

8.3.8   Fire Extinguisher

CO$_2$ is used as fire extinguisher for both consumer use and industrial applications. CO$_2$ smothers fires by preventing the contact of air (and thus oxygen) with combustible matter without damaging or contaminating materials and is used especially when water is ineffective or undesired. Noteworthy, CO$_2$ evaporates without leaving traces.

8.3.9   Water Treatment

It is worth to recall that in such application CO$_2$ is converted into its hydrated forms, which originate hydrogencarbonate anion and carbonate dianion (Fig. 6.9). Such equilibria depend on the pH of the solution and are fully reversible, in principle CO$_2$ could be recovered. The use of CO$_2$ is in quite different directions: i. Water remineralisation, ii. Industrial water treatment, iii. Well rehabilitation, iv. Water desalinization.

i. Soft surface waters or desalted waters or highly soft waters from desalination plants require remineralization in order to adjust their calcium content. In modern water works, CO$_2$ is used together with lime to raise the hardness of such soft waters to be used as drinking water. Also, at the right pH (>10), the mentioned treatment allows the formation of a protective surface layer of CaCO$_3$ in water-mains pipes (a negative fact in boilers), thus avoiding corrosion and improving water quality by slow solubilization of the deposited carbonate which solubilization is eventually fastened by adding further CO$_2$ to water. Remineralization makes these soft waters healthier and tasty. CO$_2$ addition as a way to maintaining the quality of tap water is also commonly used in Japan.

ii. Some industries produce large volumes of alkaline waste water. Iron and steel, textile and dying, pulp and paper, glass bottle washing at beverage filling plants, and many more create effluents with a pH value above 11. Such waters cannot be discharged in natural basins (lakes, rivers, sea). To lower the pH, sulphuric acid has been used for long time which requires careful handling (avoiding surplus that would produce highly aggressive acid waters) and is noxious. Carbon dioxide can be used instead for treatment of such waste water streams. Noteworthy, CO$_2$ would generate two buffer systems: H$_2$CO$_3$/HCO$_3^-$

and HCO_3^-/CO_3^{2-}, and this would prevent strong acidification of water. Overdosing CO_2 is not so dangerous as overdosing H_2SO_4; sparkling water is not dangerous at the end! Such technology, known since more than 30 years, is now growing in interest and application as people gain environmental awareness and are more conscious of the danger of storing and handling concentrated sulphuric acid. Other advantages toward the use of classical acids are use of less toxic reagents, less need of devices for monitoring, less downtime, increased human safety, longer equipment life, etc.

iii. The water-well rehabilitation technique consists in injecting gaseous carbon dioxide at the desired depth to produce an abrasive acid solution that penetrates far into the surrounding formation. Liquefied carbon dioxide is then injected under controlled conditions at various temperatures and pressures. When it comes in contact with the water, it expands rapidly, producing violent agitation and the freezing of water within the formation around the well, resulting in superior disinfection and dislodging of mineral encrustation.

After treatment the well is mechanically cured using surge/airlift methods that remove the newly dislodged particulate matter from the well. The well pump is then reinstalled and the well returned to service.

iv. The desalination of seawater to produce soft water even usable for drinking, especially in Israel and the Middle East, is more and more using CO_2 as curing agent for the reasons discussed above. Today more than 80% of the desalination plants are located in the Gulf States, and the technology is strongly expanding in Saudi Arabia. The estimated CO_2 consumption for this application is at the level of 33–50 g_{CO2} m^{-3} of desalinated water.

8.3.10   Greenhouses

The use of CO_2 in greenhouses for increasing the CO_2 level to 600 ppm instead of *ca.* 410 ppm stimulates the growth of some plants often up to 20%. This is true not only for vegetables such as tomato, cucumber, lettuce and other greens, strawberry but also for potted plants and cut flowers. This is popular in the Netherlands since long time, where LINDE's OCAP [14] has built a 85 km transport pipeline (300 km total distribution network) which distributes over 450 kt y^{-1} of CO_2 to over 550 greenhouses. Such CO_2 originates from a Shell refinery and the Alco-BioFuel bioethanol plant, both located in the Rotterdam area [15]. In the US, such application has recently started to grow (>50% in the last 10 years), pushed in the last years by the safe conditions required to grow legally cannabis.

Growing microalgae in photobioreactors (see Chap. 11) represents a way to a fast conversion of CO_2. Each kg of algae demands *ca.* 2 kg of CO_2. The economics of such practice is viable if products for food (omega 3, proteins, specific nutrients), pharmaceuticals, colors, and other similar sectors are targeted; it is not viable when fuels alone are targeted. Such application, which mimics greenhouses, is expanding as alternative to use of plants that demand arable land that in some areas is scarce.

Algae appear to be a very promising source of fine chemicals and in some cases pharmaceuticals. To size the amount of CO$_2$ used in such application is not an easy task because a variety of gas streams are used, even flue gas from power plants.

Another interesting application of carbon dioxide is its addition to irrigation water to enhance the absorption of nutrients by plants. Such beneficial effect is due to the fact that, even CO$_2$ has itself weak nutrient properties, being it an acid may mobilize some specific nutrient elements in soil that are adsorbed by roots of plants and facilitate their growth.

8.3.11   Dry Ice

Dry ice is the common name for solid CO$_2$. It is produced by injecting L-CO$_2$ into pelletizers where it converts in almost equal amounts of gaseous CO$_2$ and CO$_2$-snow, which is compressed in a cylinder and extruded or pressed at high pressure, resulting in the manufacture of various types of dry ice. Pellets and nuggets are formed when the snow is extruded through die plates, which can be interchanged to create various sizes. Dry ice is commonly manufactured in four forms: pellets (3–19 mm), nuggets (3–19 mm), blocks, and slices/slabs (see Fig. 8.3).

Dry ice immediately after its production is stored into a specially designed, well-insulated container that minimizes its sublimation and is used for storage and transportation. As dry ice sublimes easily, it is important that the production and distribution are well timed for avoiding long storage. As already discussed in Chap. 7, dry ice has a temperature of −78.5 °C and directly passes from the solid to the gas phase (sublimation) without passing through the liquid phase.

Dry ice is relatively soft (1.5–2.0 in the Mohs Scale of Hardness, second only to talc) and does not cause abrasion on most surfaces. Its density is 1.56 kg L^{-1}: for comparison that of water-ice is less than 1 kg L^{-1}. The heat of vaporization of dry ice is around 645 kJ kg^{-1} (152 kcal kg^{-1}) at 0 °C (twice the refrigerating capacity of water-ice per unit weight).

Dry ice finds a large industrial utilization in a variety of applications, almost all linked to its cooling properties. With respect to normal ice (from water) dry ice does

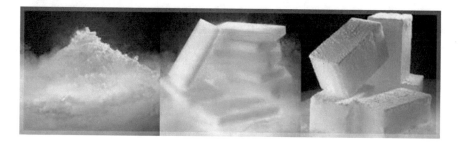

Fig. 8.3 Images of Dry ice as pellets, slices, and blocks [16]

not leave any residual liquid as it does not melt, but sublimates. Roughly 90–95% of all produced dry ice is used in cooling products during transportation. Examples are Airline catering, Food Packaging and Distribution, Home Delivery Services, Medical (Biological) Samples and Pharmaceutical Distribution, Food Processing, Winemaking (Cryomaceration), Baking, and Composite Material Manufacturing.

Another application is the *blasting* technique developed by *Cold Jet in 1986*, originally developed to remove paint and coating from airplanes in the 1970s. Such technique uses pressurized air stream to accelerate particles of dry ice on the surface to be cleaned. The particles sublimate upon impact, and lift dirt and contaminants off without damaging the underlying substrate.

The impact, with the assistance of the difference of temperature, causes detachment of dirt or paint or contaminants which are pushed away from the surface by blowing of the particle which passes to the gas phase with an increment of volume of *ca.* 800 times (Fig. 8.4). Such application works well for surface cleaning or preparation, part (not only metals but also plastic) finishing in several sectors: Aerospace, Automotive, Contact Cleaning, Composite Tool Cleaning, Electric Motor, Fire Restoration, Food and Beverage, Foundry, Oil and Gas, General Maintenance and Facilities, Historical Restoration, Medical Device Manufacturing, Mold Remediation, Packaging, Plastics, Power Generation, Printing, Rubber, Textile and Engineered Wood, Fire and Smoke Damage Removal, Mold Remediation, Water Damage, and Odor Elimination. Cold Jet was the first (1986) to file patents in these fields. Dry-ice blasting substitutes old technologies that are more aggressive and lesser safe.

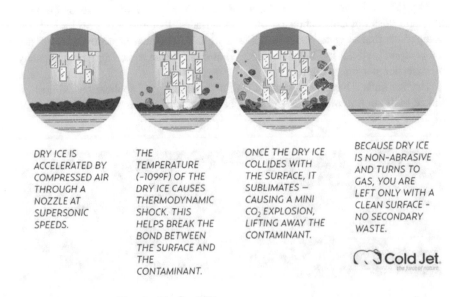

DRY ICE IS ACCELERATED BY COMPRESSED AIR THROUGH A NOZZLE AT SUPERSONIC SPEEDS.

THE TEMPERATURE (-109ºF) OF THE DRY ICE CAUSES THERMODYNAMIC SHOCK. THIS HELPS BREAK THE BOND BETWEEN THE SURFACE AND THE CONTAMINANT.

ONCE THE DRY ICE COLLIDES WITH THE SURFACE, IT SUBLIMATES – CAUSING A MINI CO$_2$ EXPLOSION, LIFTING AWAY THE CONTAMINANT.

BECAUSE DRY ICE IS NON-ABRASIVE AND TURNS TO GAS, YOU ARE LEFT ONLY WITH A CLEAN SURFACE - NO SECONDARY WASTE.

Cold Jet.
the force of nature.

Fig. 8.4 A cartoon of Dry-ice blasting [17]

8.3.12 Other Uses

CO$_2$ finds several other minor (lower amount) applications such as nuclear plants cooling, gas-operated fire arms (a Colt built in the 1890s was operated by injecting pressurized CO$_2$ to pull the bullet), serving pressure beer in bars, medical use in laser surgery, smoke effects in theatres and disco's, home-made sparkling water.

8.4 Some Economic Figures

Determining the exact value of the merchant market of CO$_2$ is quite difficult because most data are confidential. Major players in the sector are AGA, Air Liquide, Air Products and Chemicals, Inc., AIR WATER INC, Airgas, Buzwair Industrial Gases Factories, Continental Carbonic Products, Cosmo Engineering, EPC Engineering & Technologies GmbH. The global market in 2019 was estimated at 12.55 BUS$ 2019 driven by EOR that will play a key role even during next 5–10 years. It is expected to grow at a compound annual growth rate (CAGR) of 7.4% and reach US$16.76 billion by 2023 [18]. The food-grade CO$_2$ market will also experience fast growth and will play a dominant position in global food-grade gases market of 5.9 BUS$ in 2018, estimated to rise to 8.1 BUS$ in 2023 with a CAGR of 6.7%.

The sector of carbonated beverages can afford quite real figures, at least for Europe. SODASTREAM [19] commercializes apparatuses for making home-made sparkling water: its portfolio is of 300 Million €/year, mostly in Europe and especially in Germany and Northern Europe where over 2.2 million users consume over 11 million small cylinder each year. The SodaStream company has recently been acquired by PepsiCo for 3.2 BUS$.

Some trends of the market deserve comments. The German CO$_2$ industry started in 1885 with just 122 t of CO$_2$; in 1899 it was 15 000 t and in 1918 the CO$_2$ merchant market exceeded 33 000 t. In 2019, the market was around 1 Mt of CO$_2$ consumed in Germany.

In 2012, the total consumption of CO$_2$ in Europe including the eastern part (Poland, etc.) reached about 4.0 Mt/y. It is currently estimated to be above 6 Mt/y.

Excluding enhanced oil recovery and on-site uses of gaseous CO$_2$, the US merchant market is estimated to exceed 10 Mt/y (representing a value of 1.5 BUS$) and the global market could well reach 30 Mt/y.

The merchant market players (industrial gases companies as well as international suppliers of CO$_2$ production and recovery equipment) underwent consolidation in the recent years. Air Liquide has acquired Airgas, and Linde and Praxair merged to a 70 BUS$, Air Products has acquired ACP, leaving only a single company that operates alone in EU, CARBO Kohlensäure in Germany.

Mitsubishi Heavy Industries (MHI) now owns Taiyo Nippon Sanso Corp (TNSC) which itself owns Matheson, which acquired Continental Carbonic in 2014 and 4 spin-off plants from the Airgas-AL merged in the US. TNSC is also intending to acquire the European disinvestments linked to the Praxair-Linde merger.

The CO_2 Merchant Market is moving fast and moves billion dollars.

References

1. Aresta M, Dibenedetto A (2014) Catalysis for the valorization of low-value C-streams. J Braz Chem Soc 25(12):2215–2228
2. Sampson J (2018) CO_2 supply crisis. Gasworld
3. https://www.siad.com/
4. Owen-Jones J (2018) CO_2 from QUAD generation. Gasworld
5. https://www.gasworld.com/large-scale-capture-of-CO2-is-feasible-and-affordable/
6. www.NewCO2.eu
7. https://chemistry.stackexchange.com/questions/28879/what-is-the-co2-content-in-the-air-in-a-compartment-of-air-and-water-as-a-funct
8. Capuzzo A, Maffei ME, Occhipinti A (2013) Supercritical fluid extraction of plant fragrances. Molecules 18(6):7194–7238
9. (a) Dibenedetto A, Lo Noce R, Pastore C, Aresta M, Fragale C (2006) First *in vitro* use of the phenylphosphate carboxylase enzyme in SC-CO_2 for the selective carboxylation of phenol to 4-hydroxybenzoic acid. Env Chem Lett 3:145–148; (b) Aresta M ed (2003) Carbon dioxide recovery and utilization. Kluwer AP
10. Sutanto S (2014) Textile dry cleaning using carbon dioxide: process, apparatus and mechanical action, Master Thesis. Sci Chem Eng, TU Delft
11. https://www.bodan.be
12. https://linde-gas.com/
13. Van Damme F (2018) Freezing food with CO_2. Dohmeyer report
14. https://www.the-lindegroup.com
15. https://www.ocap.nl/co2-smart-grid/index.html
16. http://specco2.com/
17. https://www.a2zblasting.com/industrial-cleaning/
18. https://www.globenewswire.com/news-release/2020/02/25/1989794/0/en/2020-Carbon-Dioxide-Market-Report.html
19. https://www.sodastream.com.au/how-it-works/de
20. Aresta M, Kawi S, Karimi IA eds (2019) An economy based on carbon dioxide and water. Springer Publ., ISBN 978-3-030-15868-2

Circular Economy and Carbon Dioxide Conversion

9

Abstract

In this chapter, the relevance of Carbon Dioxide Utilization (CCU) to Innovation and Sustainability of the Chemical Industry and to the strategy of the Circular Economy is discussed. A number of examples are presented of processes which may convert CO_2 into added-value chemicals, polymeric materials, and fuels which have each a market higher than 1 Mt/y, causing a significant reduction of CO_2 emission. Only thermal processes are considered here, while solar-driven processes are discussed in Chap. 10 and the integration of Chemical Catalysis and Biotechnology is presented in Chap. 11.

9.1 The Circular Economy

Our society has increasingly adopted the "*linear*" economic model since the end of 1700s. The industrial development has pushed people to abandon the conservative attitude of the agricultural society, close to Nature circularity. Nature does not produce waste; all byproducts are recycled. In the old times, until the 1940s when plastics were not yet massively around, all daily residues were reused: fresh vegetables and garden residues were dispersed in soil, bringing back carbon and other nutrients to it, while glass and metals were recovered and reused. Residual biomass in agriculture and in wood industry were used for producing heat (used in a variety of applications, even at the manufacture level) and ashes returned to soil to which they provided useful elements or used in a variety of applications, including washing. Often in families, at night, hot-red coal was removed from fireplaces and stored in closed iron vessels where it stopped burning and made coal to be employed in future uses, instead of making ash in the fireplace. The introduction of plastics, changed many habits. The "*use once and throw away*" attitude became more and more popular. Plastics were extraneous to Nature which is unable (as we know very well

M. Aresta and A. Dibenedetto, *The Carbon Dioxide Revolution*,
https://doi.org/10.1007/978-3-030-59061-1_9

today!) to recycle them in its cycles (see Chap. 3). Since middle 1950, the use of plastics grew exponentially (today we use 360^+ Mt/y of various plastics) and the *throw-away* attitude became global reaching industries, collective services, individuals. The production of "waste," of any kind, started to grow at an unpredictable rate causing often irreversible damages to Nature: physical world, vegetation, animals, humans. "*Consumerism*" has dominated our world in last few decades, based on the false belief that the Earth resources were immense, inesaurible, enough for all. In such view, even the timespan of goods became an unnecessary attribute: short-living goods were privileged.

Misuse was encouraged, and profit became the god in all our activities. Maximizing profit at expenses of all other values has been the driver during last decades.

However, the *Linear Economy* (Fig. 9.1, top) has produced mountains of waste and damages to fragile ecosystems, reducing the biodiversity.

We need to change. We have to go back to a "*Nature inspired economy*," in all fields of our life (Table 9.1).

The "*circular or cyclic economy*," bottom part of Fig. 9.1, brings more sustainable solutions with products that are in use longer and byproducts and residues that are reused for manufacturing new goods. And such practice opens the way to new manufacturing activities, generates new work-places, makes best profit of natural resources while reducing their up-take and making them available for future generations. Interestingly, recycling is the extension of a production cycle.

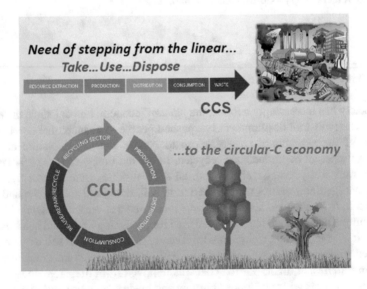

Fig. 9.1 The "*linear*" and "*cyclic*" economy

Table 9.1 Comparison of *Linear* and *Circular* Economy. The Table can be read horizontally (through a row) and vertically (through a column)

Linear Economy	Pros Cons	Circular Economy	Pros Cons
Natural resources are continuously extracted	They are used one single time. Are not replinished	Natural resources are extracted and used but….	The extracted materials have several lifes
Residues of processes are considered as waste	Goods have a single short life	Residues are recycled	Recycled residues are secondary raw materials
Waste is disposed	Disposal sites are scarce and leaching of toxic substances pollutes the environment	Residues are not considered waste. Waste is reduced, or even not produced	A cascade of technologies can be exploited for stepwise downgrade use
For disposal, waste needs to be transported to the suited site	Transport of watse may generate a strong environmental impact	Residues may, may not need transportation	Reuse of secondary raw materials can be set up at the same industrial site where they are produced: clustering of processes can avoid long distance transport.
Disposal of waste is a cost	In general, there is no benefit from disposal	Secondary raw materials have a value	There is a profit from recycling goods
Disposal sites may leach toxic compounds	Leaching may pollute water, soil and air, reaching plants, animals and humans	A final end-of-lives waste is anyway produced, in general this can be inert.	Finally a non-usable waste is produced. Disposal of such waste has a cost but the amount is much lower.

The extraction phase is replaced by the dismantling of the used goods, that in a sense is a kind of mining from a different environment than Nature, maintaining the same complexity and the same issues of working natural raw materials, if not representing a new challenge, because of the higher level of complexity. This is the case of recovering all single elements that are present in a handphone, which can contain up to some 28 different elements [1]. Such a matrix is not at all simpler than a natural ore! Mining single elements from billions handphones (there are more handphones than people on the Earth!) means to recover tonnes of different elements. One billion handphones contains, just to cite the most abundant: 15 000 t_{Cu}, 3 000 t_{Al}, 3 000 t_{Fe}, 2 000 t_{Ni}, 1 000 t_{Sn}, 100 t_{Au}, 20 $t_{Pd,In,Ta}$.

Figure 9.2 correlates *Value and Timespan* of goods in a Linear or Circular Economy frame. Single use (A) produces a single time profit for a short life product. Recycling can generate profit n-times for the same product that has several

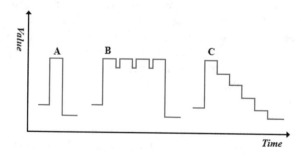

Fig. 9.2 Value and lifetime of raw materials according to the Linear Economy (**a**) and the Circular Economy (**b** and **c** represent the case of recycling to the same function or to lower quality products, respectively)

lives (B) or for a cascade of products (C) of decreasing value. The (B) case is relevant to those products that can play the same function n-times: this is the case of metals and some colored glasses (white glass requires the sorting of colored glasses), and carbon as well.

Case (C) is relevant to those goods that with recycling do not recover the original properties (this is the case of plastics or paper: plastic used for food packaging will not recover the required properties and will be used for making goods with lesser purity requirements; recycled document white paper will be used for making common writing paper, or newspaper paper or cardboards because recycled fibers do not have the resistance and quality of the original material, neither the same white color). Water can be in principle recycled n-times, but for reaching the grade of drinking water requires numerous treatments that must guarantee the absence of toxic products and dangerous microorganisms. Most recycled water is used for watering or for domestic (non-drinking) and industrial uses.

In several cases, recovery of goods and recycling is less expensive (economically and energetically) than using fresh raw materials. Therefore, recovery and recycling of goods meet the sustainability principles, extend the availability of natural resources, produce less waste, lower the environmental burden of our anthropic activities and may sustain the development of our society without (or with much less) damages to our Planet.

The circular economy will bring our society attitude back by a few decades to the style of the agricultural society: a used good is not a waste, it is a source of materials that still can be used. Used goods are not dead bodies to throw away, they are precious donors of working parts. The most difficult part of the job is now to educate people to change their attitude from consumerism to saving and having care of goods, to move from *use-once-and-throw-away* to *use-take care-store-reuse*, to consider any good as long-lasting more than just made for one or few days life: a longer life is a work opportunity for all those active in the maintenance sector. If we want to *save resources*, we have to *target longevity of produced goods* and re-circulation of their usable parts at the end of their life. This new attitude is already producing changes in our world. As an example, we consider worthwhile reporting the change that is going on in the IT market which must obey a number of "*sustainability oriented*" criteria that cover all phases of the product life cycle: material sourcing and manufacturing; use and reuse; recovery and recycling. Such criteria impose that (i) IT products must have replaceable batteries, to avoid that a good is disposed just because the battery is off; (ii) reusability of devices must be assured; (iii) spare parts must be available even three years after the product is off production; (iv) products must be upgradable; (v) parts of a device must be recyclable and interchangeable; (vi) used products must be returned to the factory instead of being disposed. These directions go opposite the one we have been stepping through until now!

As discussed in Chap. 1, recycling of water, metals, plastics, paper, and cardboard has become more and more popular with time. Some goods (metals) are recovered at interesting rates (50$^+$%), other much less. Recycling carbon is actually around 230 Mt/y, a large value per se (it is practiced at the industrial level since

1860 s) but still very marginal with respect to the emitted CO_2 (35 000 Mt/y): it is very urgent to enlarge its exploitation. *If we wish to develop a C-neutral society, we have to learn from Nature, which has developed in million years a perfect balance between the used and produced CO_2.* Such balance cannot be reached by disposing CO_2, but by reusing carbon. The capture of CO_2 from point sources (Chap. 6) is a mature technology, while the direct capture from the atmosphere requires further investigation for reducing costs and developing materials able to better uptake CO_2 at the low concentration typical of air. Table 9.2 highlights the differences between CCS and CCU.

Table 9.2 Highlights of CCU and CCS. *Reproduced with permission from Ref. [6b], Copyright 2019. Springer*

CCS Strength	Issues	Property under Scrutiny	CCU Strength	Issues
TRL=8-9		*TRL*	**TRL=4-9**	
Injection technologies are well known. Some 5Mt/y are disposed Sites are characterized by a different permanence Transportation is practiced on a large scale in industrial areas.	Sites are not ubiquitous. Presently, disposal is mainly for EOR. Doubts exist about the permanence and diffusion in new sites.	*Ubiquity* *Amount used* *Permanence*	A CO_2 conversion-plant can be built in any place 220 Mt/y of CO_2 are used. Some processes are old of 150 years (urea, aspirine, inorganic carbonates).	When used, compounds give back CO_2. Only polymers may have a sequestration time of decades.
	Safety of pipelines in civil areas is a bottleneck.	*Safety*	CO_2 is not toxic. It represents the ideal substitute of phosgene or CO	CCU is not for storing CO_2 but for cycling-C
CCS is for long time disposal. Better knowledge of disposal sites may increase safety	CCS is apt to Linear Economy and produces additional CO_2 during storage	*Expected Innovation*	CCU is Circular Economy: it is ideally used coupled in a solar-wind-geo (SWG) energy frame	Value chains and Clustering processes may optimize CCU
CCS has a sense in a fossil-C-based energy frame. It makes sense in the very short time and will lose power in future	CCS fails to be implemented in a frame of large scale SWHG exploitation	*Fossil-C vs Perennial SWHG energy frame*	CCU can turn on air-CO_2 without fossil-C extraction: emitted CO_2 can be recycled into chemicals, materials, fuels continuously.	Formation of CO_2 is much faster than some fixation processes
CCS is said to be able to store all CO_2 generated in the combustion of C-based materials on the Earth	Disposal of CO_2 is not zero CO_2-generation The real potential must be verified.	*Perspective amount of CO_2 avoided*	7-10 Gt/y of CO_2 can be converted into a variety of compounds and fuels, in a frame of SWG exploitation. CCU is the future	Integration of technologies and system approach is necessary
CCS needs energy for shipping and housing CO_2. Energy cost depends on the distance btw source-disposal	A loss of usable energy is the effect: from 25 to 50+%	*Energy requirement*	Reactions of CO_2 with electron-rich species are exoergonic. Reduction reactions are endoergonic	Making fuels needs hydrogen that must come from water
Popular acceptance of CCS has been randomly enquired. Attitude of people must be deciphered through larger enquiries.	A lot of information must be delivered to common people before their response to accept/reject CCS can be sound.	*Popular acceptance*	CO_2 is in the air and used by Nature for making million products: this is under the eyes of everybody. The industrial exploitation needs to be illustrated.	Information still needs to be spread around for clarifying basic concepts (CO vs CO_2!)
As mentioned above, the storage technology is well known with spent gas-wells. The injection technology is well known by oil-gas industry. The knowledge of new sites must be developed to high level before use.	Disposal sites require still a lot of investigation for knowing their long-term storage, leakage and diffusion. This is a prerequisite for safety.	*Research Needed*	The knowledge of the reactivity of CO_2 under different conditions is quite good, even if new catalysts can be discovered and new processes that may allow to implement a larger scale conversion	Coprocessing of CO_2 and water under solar irradiation is a must for the future avoiding hydrogen production

The recovered CO_2 can be used in chemical reactions and converted into a variety of useful compounds (chemicals, materials, fuels) that have a value. The utilization of CO_2 as building block, comonomer for polymers, or source of carbon for fuels generates profit and can pay for the capture and conversion costs. CCU is, thus, quite different from CCS also from the economic point of view. A comparison of both is reported in Table 9.2.

Before analyzing the various uses of CO_2, it is opportune to set a number of rules (Window 9.1A-W9.1A) and clarify some basic concepts (Window 9.1B-W9.1B) relevant to CO_2 conversion. W9.1A lists a number of "*must*" for CO_2 use: in simple words, any process we wish to exploit for converting CO_2 cannot produce more CO_2 than it converts and must reduce the environmental burden with respect to processes on stream not based on CO_2. These are very stringent rules that impose the necessity of a selection of opportunities. A bonus would come from engineering industrial sites and clustering processes, so that a cascade of uses is implemented among a number of different processes for a better exploitation of resources and new CO_2-conversion processes are clustered with processes in which CO_2, and other secondary raw materials, are formed, avoiding long distance transport. W9.1B summarizes the possible processes in which CO_2 is involved, concerning its C-oxidation state (+4, the reader can see Appendix A for the definition of *oxidation state*) that can remain the same or be reduced to lower values. In the latter case, energy and hydrogen are required and this demands a deep change with respect to the current energy frame.

W9.1A.   Musts for CO₂-use

- Any CO_2-based process must release less CO_2 than it converts.
- Any innovative process must emit less CO_2 than the relevant process on stream.
- The new process must minimize the energy input and maximize the *Energy Ratio*: *Energy out/Energy in*.
- The new process must be assessed by *LCA* that should demonstrate not only a reduced C-Footprint, but also a minimization of the impact on other categories (at least: *Acidification potential-AP; Eutrophication potential-EP; Global warming potential-GWP; Air pollution-AP*).
- H_2 cannot be generated from fossil-C.
- Whenever possible perennial energies (Solar-, Wind-, Hydro-, Geo-power) must be used.
- Integration of Biotechnologies and Catalysis must be implemented.
- Clustering of Industrial Processes is essential for a better recovery-recycling practice.

W9.1B. Key concepts about CO₂-use

- CO_2 can be incorporated as such into an inorganic or organic compound: such process can be exothermic or slightly endothermic. This process can be carried out using fossil-C. Heat can be recovered from exothermic processes.
- CO_2 can be incorporated as a reduced moiety (CO) into an organic compound: such process requires energy (mild to high temperature) and an oxo-phyle to fix the excess O-atom. For low-energy processes the C-based energy frame is suited, most likely non-fossil energy is required. In any case oxo-phyle must be recoverable and recyclable to avoid organic waste production.
- CO_2 can be reduced to hydrocarbons of general formula C_nH_{2n+2}, from methane to liquid hydrocarbons. In this case hydrogen is necessary, that must come from water or waste organics, not from fossil-C.
- CO_2 can be stored into very long time span inorganic carbonates or moderately long living polymers.
- Processes must be selective and avoid the production of organic waste that generates CO_2.

Items listed in W9.1A&B represents the frame to the matter discussed in this and following Chapters. The Carbon Dioxide Capture and Utilization (CCU) has strict rules that must be obeyed otherwise the risk is that more CO_2 is produced than used and a larger environmental impact is generated instead of helping our Planet. In order to avoid disasters, new processes must be assessed with the help of the Life

Cycle Analysis (LCA) Methodology. In applying such a tool, correct parameters must be used as LCA is an interesting comparative tool that should be used with special care, being it very sensitive to fed data which must be *real* process data, collected with much attention. An absolute use of LCA is quite complex, if not impossible.

Moreover, LCA is changing [2] and increasing its complexity, as it is moving from single product-process to cluster of processes and cascades of products assessment, including waste management.

Previous Chapters have shown that CO_2 is a peculiar molecule, which lays in a deep energy well and only a limited number of reactions are allowed that are energetically favored. Figure 9.3 represents the energy constraints to CO_2 conversion.

However, we shall proceed from now on by following a path along which the reaction energy will increase: moving from durable materials and other compounds in which CO_2 can be stored as such, to intermediates and fuels a *crescendo* of energy demand will be encountered, with the need of adopting different technologies and using different energy sources, and eventually even using non-fossil hydrogen. As anticipated in W9.1B, the chemical utilization of CO_2 can be categorized into three main classes of reactions [3].

A. *Incorporation into chemicals in which the entire-CO_2-entity is preserved, without C–C bond formation*

In such case, the oxidation state of the C-atom remains equal to +4. Such processes can be low in energy requirement or even be exoergonic. The CO_2 molecule easily reacts with electron-rich (energy-rich) co-reagents such as amines-RR′R″N, or O-containing species: O^{2-}, OH^-, OR^-, or similar systems. Typical products are

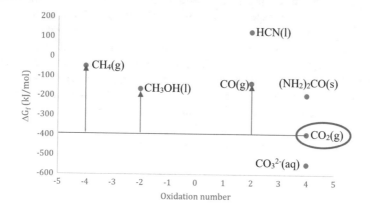

Fig. 9.3 Gibbs-Free Energy ($\Delta G°_f$) of CO_2 and some other C1-Molecules. Organic carbonates and carbamates lay within the red-circle, inorganic carbonates (CO_3^{2-}) are lower in energy, all other species are higher in energy. In order to convert CO_2 in any species above it, energy must be provided to CO_2, represented by the red segment in the Figure. Often, hydrogen is also needed

inorganic carbonates $[M(M'_2)CO_3]$, carbamates (the ammonium salt $HRN–CO_2^-$ ^+H_3RN, or the $HRN–CO_2R'$ species, both derivatives of the labile carbamic acid $HRN–CO_2H$), organic carbonates $[(RO)_2C=O$, putative esters of carbonic acid, $(HO)_2C=O]$, hydrogencarbonates $(MHCO_3)$. In this class of products fall also polycarbonates obtained by co-polymerization of CO_2 with epoxides.

B. *Carboxylation of organic substrates*

In the carboxylation of organic substrates, a C–C bond is formed and the oxidation state of C goes down to +3 from +4. Examples are (i) reactions in which CO_2 reacts with energy-rich molecules such as olefins, alkynes, dienes (conjugated and cumulated) to afford specialty chemicals such as lactones, pyrones, esters, among others; (ii) insertion of CO_2 into C–H bonds with formation of acids (Eq. 9.1).

$$R–H + CO_2 \rightarrow RCOOH \qquad\qquad (9.1)$$

The former reactions occur under quite mild conditions and can be exploited in an energy system based on fossil-C, the latter (Eq. 9.1) are quite common in Nature and are of great interest for Industry as their implementation would reduce the C-footprint and environmental impact of chemicals such as organic acids, as will be discussed in next paragraphs.

C. *Reduction to energy products*

This class of reactions includes the conversion into chemicals in which the C-atom has a lower oxidation state than +3: from +2 (e.g., CO) to −4 (e.g., CH_4). The energy content of the resulting chemical (and, thus, the energy to be delivered to CO_2 for its conversion) depends on a series of factors, among which the oxidation state of the C-atom in end species, the number of C–H bonds formed, the number of C–C bond formed.

Table 9.3 shows the thermodynamic properties of some molecules derived from CO_2 upon hydrogenation. Data show that there is not a linear relationship between the change of the oxidation state with respect to CO_2 (Column 3) and the change of Free Gibbs energy (Column 7) moving from CO_2 to CH_4. A peculiar situation is represented by methanol and ethene in which the C-atom presents an oxidation state equal to-2, but the two compounds present different energy content per C-atom due to the different environments. Let us now consider some chemical uses of CO_2.

Table 9.3 Thermodynamic properties [4] of some molecules derived from CO_2 upon hydrogenation. *Reproduced with permission from Ref. [6b], Copyright 2019. Springer*

Compound	Formal oxidation state of C	$\Delta_{Ox} = n_{ox,X}\text{-}n_{ox,CO_2}$	$\Delta H°$ kJ/mol	$\Delta S°$ J/mol	$\Delta G°$ kJ/mol	$\Delta G°$ kJ/mol C	#
CO_2	+4	0	−393.51	213.6	−457.2	−457.2	1
CO	+2	−2	−110.53	197.6	−169.4	−169.4	2
CH_2O	0	−4					3
C_2H_2	−1	−5	226.7	200.8	166.9	88.45	4
CH_3OH	−2	−6	−238.7	126.8	−276.5	−276.5	5
C_2H_4	−2	−6	52.3	219.5	−13.1	−6.55	6
C_3H_8	−2.7	−6.7	−103.8	269.9	−184.2	−61.4	7
C_2H_6	−3	−7	−84.7	229.5	−153	−76.5	8
CH_4	−4	−8	−74.8	186.2	−130.3	−130.3	9

9.2 Carbon Dioxide Conversion (CCU)

What contribution can CCU give to reducing the putative impact of CO_2 on Climate Change, how can it improve our life, what is the time to its large-scale exploitation, what will be its intensity, what technologies must be developed? In this and next chapters an answer to such questions will be strived.

9.2.1 CCU and Resources Conservation

The utilization of CO_2 as building block or source of carbon means recycling carbon, mimicking Nature. This is a virtuous practice that reduces the transfer of fossil-C to the atmosphere, supports the sustainability of the chemical industry (and transport sector), and preserves fossil-C resources for next generations. Summarizing data reported in Chaps. 1–3, the consumption of fossil-C today ranges around 11.5 [5a]–12.5 Gtoe/y [5b] with a forecast of reaching 25 Gtoe in 2035. In a Business as Usual (BAU), the CO_2 emission will rise from 37 to over 70 Gt/y! The daunting figure is that the estimated fossil-C reserves (coal, oil, and gas) are today quantified at 941 Gtoe. According to the BP forecast, [6] at the actual rate of consumption of fossil-C, humanity have crude oil available until year 2066, natural gas until year 2068, and coal until year 2169.

Such figures are impressive! We hope that new reserves are discovered or deeper sites not reached so far will be exploited. This is a wish, what is the reality?

As for today, the perspective availability of fossil-C of *ca.* 150 years asks for immediate action and use of renewable-C (and CO_2 is renewable carbon), for making chemicals and fuels, mimicking Nature; this is an ethical must with scientific or technical issues. However, one could ask at what cost recycling can be done and whether or not recycling carbon is sustainable and its extension. Figure 9.4 presents the history of the sources of energy and goods for humans (food is not considered). Until *ca.* 1750, biomass was practically the only real source of energy and goods (materials, fibers). With the industrial revolution and the exploitation of coal-mines first, and oil- and gas-fields later, fossil carbon has increased its role as source of energy and goods competing and winning over biomass because of the higher intensity of energy and different kind of constituents, which have a much lower content of oxygen with respect to biomass.

Since the beginning of the last century, humans have built their development and the satisfaction of their needs on coal chemistry and petrochemistry. Fossil-C has been the source of most goods we use for living: materials such as polymers, fibers for clothes, fuels, pharmaceuticals, cosmetics, body-care products, and leisure products. Today, we are moving toward the use of renewable-C, aiming at reducing the use of fossil-C and its transfer to the atmosphere. CO_2 is the form used in Nature for cycling carbon. As already discussed above, biomass alone cannot satisfy our needs and substitute fossil-C. (see also Chaps. 10 and 11) Therefore, the man-made utilization of CO_2 can integrate biomass and the natural C-cycle in offering a contribution to reduce the impact on climate of anthropogenic activities. CO_2 can be used in many different ways: as comonomer for polymers, building block for chemicals, source of carbon for fuels. The latter option requires hydrogen that must be produced from water at an economically viable price. The price of H_2 from methane reforming (MR) is around 1.3–1.5 US$/kg today, while the price of PV-H_2 (hydrogen generated through electrolysis of water using photovoltaic electric energy) is two to three times higher.

Biomass

Biomass

Biomass

Biomass

Biomass

<1750	1750 ++	Last two centuries	Our days	Future

Fig. 9.4 Sources of energy and goods in the human history: past and future *Reproduced by permission from Ref. [6b], Copyright 2019*

It is foreseen that by 2040, the price of PV-H$_2$ may match that of MR-H$_2$ because the cost of PV devices will fall down while their lifetime and efficiency will increase. This high cost of non-fossil H$_2$ explains why the conversion of CO$_2$ into fuels has not been pursued with the due intensity so far; the cost of produced fuels would have been so high that they would have not had market. But the future is different! Installed PV is increasing at a high rate: from 51 GW in 2015, to 305 in 2017, 969 in 2025 and 3 500–4 500 GW [7] by 2040, while the cost of PV-materials will decrease by 50%. Additionally, the direct use of solar energy for the efficient concomitant CO$_2$ reduction and water oxidation has the possibility of being developed with efficiency higher than 10% in coming years [8].

A new paradigm for CO$_2$ conversion [9] is in front of us: the *CO$_2$ Revolution!* Let us make an analysis of options we have for CO$_2$ conversion, moving along the energy rising slope.

9.2.2 CCU, Sustainability and Innovation

It is worth emphasizing that if the *energy industry can be decarbonized, the chemical and polymer industry cannot.* One of the most exciting features of CCU is its ability to *conjugate sustainability and innovation.* The use of CO$_2$ is per se a new attitude that may produce

- *more direct reactions* with *reduction of reaction steps* and, thus, of *waste production*;
- *less use of fossil-C and general resource saving*;
- *lower process and separation energy*, coupled to other *tangible benefits* such as
  - *safer working conditions*,
  - *raw material diversification*,
  - *reduction of carbon footprint*,
  - *lower overall environmental impact* (reduction of not only GHG emission, but also reduced burden on other categories such as human toxicity, soil and water toxicology, air pollution, acidification, etc.),
  - *lower overall industrial waste production* (with subsequent lower CO$_2$ emission in waste treatment) and *less toxic*.

Innovative processes based on CO$_2$ are those which while reducing the C-footprint of a good, do not increase the impact on other environmental categories (see W9.1A). CO$_2$ reduction must be coupled to the reduction of waste and emissions, which can negatively impact other environmental categories. Reducing CCP, while increasing the impact on other categories (see Appendix F) is not wise as it does not represent a solution to the problem but means aggravating the environmental burden. CCU must contribute to improving the state of our planet. In the synthetic Chemical Industry, CO$_2$ represents the safer alternative to either phosgene (COCl$_2$, a toxic species, LC$_{50}$ = 3 ppm) [10] or CO (poisonous gas) [11]. Implementing safer conditions means reducing Capital Expenditure (CAPEX), as a lower investment in safety measures will be necessary, including the use of special materials, assuring longer life to plants with lower Operative Expenditure (OPEX) (less corrosion of plants). Safety comes at expenses of reactivity. Phosgene is very reactive and allows reaction going at room temperature, but generates difficult waste and requires that plants are insulated in a dome linked to air treatment plants [12]. Transport of phosgene is banned and even the use of phosgene is prohibited in several countries, after some serious cases of human poisoning. Therefore, new synthetic methodologies are welcome and meet the industrial interest and CO_2 is the ideal candidate in some applications. CO_2 can be the origin of CO (see DI 2.1) through the Reverse Water Gas Shift reaction, which requires non-fossil H_2 (Eq. 9.2) whose production from fossil-C emits CO_2. (see Chap. 2).

$$CO_2 + H_2 \rightarrow CO + H_2O \,(RWGS) \tag{9.2}$$

Therefore, while the substitution of $COCl_2$ with CO_2 would be possible as soon as the right technologies are ready, the substitution of CO requires that PV-H_2 will have a cost comparable to MR-H_2 or coprocessing of H_2O–CO_2 may produce CO, an even more difficult process.

9.2.3 CCU and the Sustainability of the Polymer Industry

CO_2 can be used as co-monomer in the production of polycarbonates-PC which are made of alternate organic moieties and CO_2 (W9.2). The PC market is sized at ca. 4.5 Mt/y and such materials find a wide application in several fields, from medical equipment, to CDs and DVDs, electronic industry, internal components of cars, and the construction of tops and separation walls of buildings (Fig. 9.5).

Fig. 9.5 Utilization of polycarbonates in buildings

W9.2 The polycarbonate structure

Polycarbonates are polymers which have organic functional groups linked together by carbonate groups. The most used organic moieties are the propene, styrene, and bis-phenolA. Propene and styrene polycarbonates can be made by reaction of propene- and styrene-oxide with CO_2. Viceversa, Bisphenol-A carbonate is still made from phosgene.

Depending on the nature of the organic moiety, the regular alternate insertion and the length of the polymeric chain that will determine the "molecular mass," polycarbonates may present various degrees of flexibility-rigidity-hardness. The synthesis of polycarbonates from CO_2 is based on its direct reaction with epoxides (such as propene- and styrene-epoxide, see W9.2). Bisphenol A-carbonate cannot be prepared in this way. In order to use CO_2, an indirect route must be followed, based on the preliminary synthesis of an alkyl carbonate such as DMC (DiMethylCarbonate) which is then reacted with Bisphenol-A (W9.2 lower part) affording the polymer and giving off methanol. Several companies (among which Covestro in Europe and Novomer in USA predominate) are now developing such synthetic methodologies that are much safer and produce less waste (lower environmental impact) than the old one based on phosgene. The benefit in terms of reduced emission of CO_2 is given by a reduction factor of 5 with respect to the phosgene technology, or even higher. Polycarbonates represent the frontier field in new applications of CO_2.

Polyurethanes, characterized by the $-N-C(O)O-$ moiety, represent another class of polymers potentially derived from CO_2 of great industrial interest. Polyurethanes are today made from phosgene via isocyanates (R-NCO) (Eq. 9.3) and find wide application as foams (acoustic and thermal insulators), in packaging, and in the fabrication of mattresses and a number of other goods.

Di-isocyanate   **Polyol**   **Polyurethane**

$$(9.3)$$

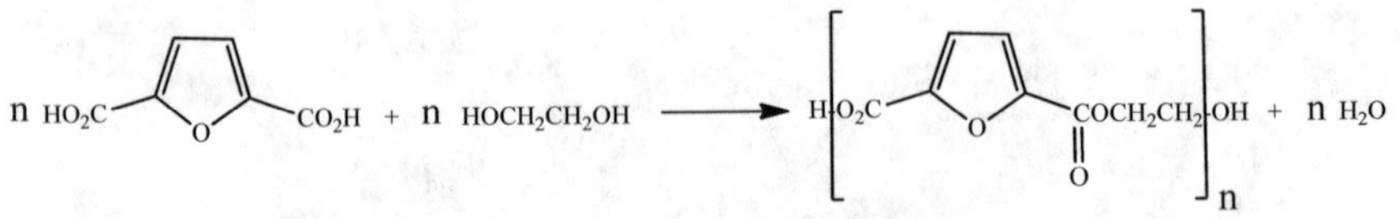

$$n\ HO_2C \underset{O}{\diagup\!\!\!\diagdown} CO_2H\ +\ n\ HOCH_2CH_2OH\ \longrightarrow\ \left[H O_2C \underset{O}{\diagup\!\!\!\diagdown} COCH_2CH_2 OH \right]_n\ +\ n\ H_2O$$

Fig. 9.6 Synthesis of polyfuranoates: All –C– Derived from CO_2

An alternative route to isocyanates which are believed to be carcinogenic is the reaction of amines with organic carbonates, or the reaction of amines, CO_2, and alcohols.

As mentioned above, polymers containing CO_2 can be considered a way to store CO_2 for a few decades: CO_2 is given off when polymers are burned or destroyed. An appealing route is making polymers which have all C-atoms derived from CO_2. This is the case of poly-furanoates that can substitute terephthalates (polyesters derived from fossil-C) (Fig. 9.6). The integration of the use of biomass-derived monomers (furandicarboxylic acid-FDCA, left part of Fig. 9.6) and synthetic chemistry/catalysis allows to produce *all-C-from-CO₂ polymers*. In fact, diols ($HOCH_2CH_2OH$) can be electrochemically derived from CO_2, while FDCA is made from biomass which is produced from CO_2.

The examples above are low energy processes which use CO_2 in a direct or indirect way and can be exploited in the short–medium term, even within an energy system based on fossil-C.

9.2.4 CCU and the Innovation in the Synthesis of Chemicals

Carbon is essential for life. Human body is made of 18–20% carbon. The absolute majority of goods we use daily contains carbon and the majority of them (85–90%) are produced via catalytic processes, with a minor share of natural products. However, carbon will always be around even in the far future; we can imagine change in its source, but we cannot substitute it. And the change of source is a big one: form fossil-C to CO_2. The aspects that make the CO_2-based chemistry economically and environmentally appealing, while empowering sustainable processes, will be now examined. The aspects relevant to the use of CO_2 as *building block* will be discussed here, while fuels will be treated in the next paragraph. Only chemicals with a market close to or higher than ca. 1 Mt/y will be considered.

9.2.4.1 Formic Acid, HCO$_2$

Market of *ca.* 800 kt/y and annual growth estimated at 4.95% by 2027 [13].

Has a utilization in sectors such as animal feed, rubber and leather production, dyeing and finishing textiles, cleaning agents, goods preservatives. The use as H_2-carrier/storage may increase its market to tens Mt/y. Formic acid is currently made through routes based on CO, produced *via* coal or methane reforming that cause emission of CO_2. Its direct synthesis from CO_2 and H_2 (Eq. 9.4a) is attempted since long time as would be highly beneficial (reduction of CO_2 emission). Reaction 9.4a has some thermodynamic limitations ($\Delta G° = +6$ kcal/mol) due to the negative entropic contribution (two gases are converted into a liquid). The equilibrium concentration is low. If carried out in presence of a proton acceptor (amine or water/Na(K)OH) the thermodynamics is much improved ($\Delta G° = ca. -30$ kcal/mol) and the reaction is shifted to the right, but producing salts more than the free acid (Eq. 9.4b). The latter can be obtained by treating the salts with a strong acid (Eq. 9.4c), increasing the complexity of the process and waste production.

$$H_{2g} + CO_{2g} \rightarrow HCO_2H_l \qquad (9.4a)$$

$$H_{2g} + CO_{2g} + RR'NH[Na(K)OH] \rightarrow RR'NH_2^{+\,-}HCO_2\left[Na(K)^{+\,-}HCO_2 + H_2O\right] \qquad (9.4b)$$

$$RR'NH_2^{+\,-}HCO_2 + HX \rightarrow HCO_2H + RR'NH_2^{+\,-}X \qquad (9.4c)$$

Noteworthy, some catalysts have been developed so far for the synthesis of formates (Eq. 9.4b) which could be considered for industrial application [14]. Should the ongoing attempts to their use in the synthesis of the free acid, instead of salts, succeed (by using SC-CO$_2$ [15] or water as reaction solvents [16]), the direct synthesis of formic acid from CO_2 and water (as source of H_2) will be ready for exploitation. The interest in formic acid lays also in the fact that reaction 9.4a can be easily catalytically reversed to CO_2 and H_2 making, thus, formic acid an interesting liquid carrier (or storage) of hydrogen for fuel cells.

9.2.4.2 Carboxylic Acids, R–CO$_2$H

$$\underset{R\quad\qquad OH}{\overset{\displaystyle O}{\underset{\big|}{C}}}$$

R=aliphatic or aromatic group. When R=CH$_3$, acetic acid is obtained

Acetic acid has a market of is *ca.* 15 Mt/y. It is made by either microbial oxidation of bio-ethanol (vinegar) or synthetized from CH$_3$OH and CO. The market of *long chain*(C10–C18) carboxylic acids is over 10 Mt/y. The overall value was 13 BUS$ in 2017 with a grow of 5%/y.

Aliphatic and aromatic acids find a large industrial use.

Butyric (C4), valeric (C5), palmitic (C16), and stearic (C18) aliphatic acids are extensively used in the manufacturing processes of soaps, detergents, cosmetics, cleansing agents, and other personal care products [17].

Aromatic acids such as benzoic acid (C6) and its derivatives, phthalic and terephthalic acids (C8) find industrial utilization in the field of polymers. Acids are synthetized either through routes based on CO or via oxidative processes or hydrolysis of cyanides that either have a low Carbon Utilization Fraction-CUF (Eq. 9.5a) or may have a heavy environmental impact (Eq. 9.5b).

$$(9.5a)$$

$$R{-}CN + 2H_2O \rightarrow R{-}CO_2H + NH_3 \qquad (9.5b)$$

The direct synthesis based on CO$_2$ (Eq. 9.6) is highly wished, but demands the C–H activation, a not easy process to be carried out by thermal catalytic routes, maybe easier if performed by photochemical catalysis [18] (see Chap. 10). Noteworthy, reaction 9.6 has a CFU = 1 and does not produce toxic waste.

$$R(Ar){-}H + CO_2 \rightarrow R(Ar){-}CO_2H \qquad (9.6)$$

Direct carboxylations are quite appealing for the high level of innovation and the environmental benefits due to quasi-zero organic waste production and significant reduction of CO$_2$ emission.

The production of long-chain carboxylic acids from hydrocarbons and CO$_2$ would represent an interesting route to biodegradable surfactants (that would substitute on the market the products used today, i.e., benzene sulphonates, not easily biodegraded) with great environmental benefit.

9.2.4.3 Acrylic Acids, $CH_2=C(R)CO_2H$

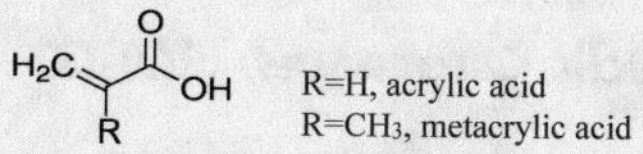

R=H, acrylic acid
R=CH₃, metacrylic acid

R=H, acrylic acid
R=CH₃, metacrylic acid

The overall market of acrylic products was *ca.* 6.6 Mt/y (2017) with a perspective annual growth of 6.3%/y and an estimated global value of 9.88 BUS\$ by 2022.

The market of acrylic acids derivatives [19a] is built up by several different compounds, such as

Acrylic Esters (Methyl Acrylate, Ethyl Acrylate, Butyl Acrylate, Ethyhexyl Acrylate), Acrylic Polymers (Acrylic Elastomers, Super Absorbent Polymers, Water Treatment Polymers, Ammonium Polyacrylate, Cyanopolyacrylate), End Use (Diapers, Surface Coatings, Adhesives and Sealants, Plastic Additives, and Water Treatment).

The production of acrylic acid is today based on the non-selective oxidation of propane to propene and of the latter to acrylic acid, (Eq. 9.7) a route that has a low CUF and produces a lot of organic waste and CO_2 due to the difficulty of driving very selective oxidation processes.

$$CH_3CH_2CH_3 \rightarrow CH_3CH=CH_2 \rightarrow HO_2CCH=CH_2 \tag{9.7}$$

$$CH_2=CH_2 + CO_2 \rightarrow CH_2=CH-CO_2H \tag{9.8}$$

The direct reaction of ethene with CO_2 (Eq. 9.8) is feasible, even if it has several thermodynamic and kinetic bottlenecks. The coupling is promoted by metal (Ni, Mo, W, Pd) complexes in a low or zero oxidation state with production of a metallacycle (**A**, Scheme 9.1) which is then converted into a hydride-acrylate species (**B**, Scheme 9.1). Reductive elimination should give back the catalyst and acrylic acid.

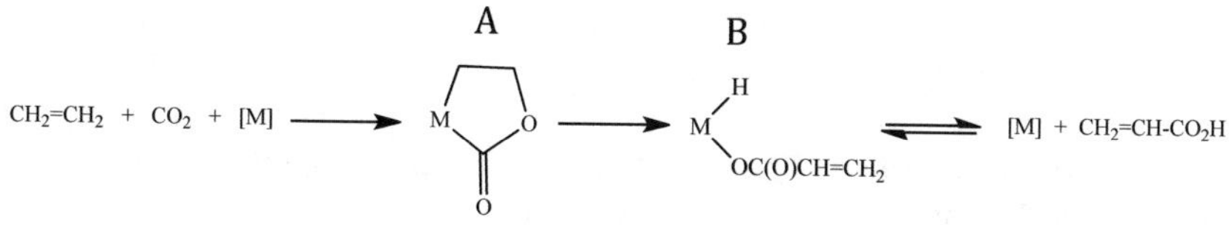

Scheme 9.1 Pathway to the formation of acrylic acid from ethene and CO_2

Such approach is still in its infancy as Turn Over Number-TON of 10–100 have only recently been reached using Pd [19b–f]. Interestingly, ethene can be made from CO_2 via electrolysis so that all carbons of the acrylic moiety could be derived from CO_2.

9.2.4.4 Linear and Cyclic Carbonates, $(RO)_2CO$

Linear carbonates   Cyclic carbonates

Organic carbonates find several industrial applications as solvents, reagents, co-monomers for polymers. Their market is of very few Mt/y, but it should be considered that most of their production does not reach the market as it is for captive use in the polycarbonate industry (>4.5 Mt/y) or other applications.

The utilization of CO_2 in this field has a great potential. The technology is at high TRL (see Appendix D) for making cyclic carbonates from epoxides and CO_2 but is not yet mature for the direct carboxylation of alcohols or diols with CO_2.

Linear carbonates can be produced by direct carboxylation of alcohols (Eq. 9.9). *Bottlenecks* are the *thermodynamics* of the reaction (that causes an equilibrium concentration of 1–2% of the carbonates that cannot be efficiently distilled out) and the *slow kinetics*.

$$2\, ROH + CO_2 = (RO)_2CO + H_2O \tag{9.9}$$

However, reaction 9.9 is slightly endergonic, [20] favored by a temperature close to 300 K and water elimination. Active and selective catalysts must be developed, with the help of DFT calculations, [21] to keep low the complexity of the reaction system and the energy necessary in the separation process [22]. Water elimination has been attempted using inorganic or organic water traps and membranes [23]. The latter option avoids the regeneration of the water traps and needs the development of new robust and selective materials.

Cyclic carbonates are produced mainly by direct carboxylation of epoxides [24] (Eq. 9.10) or carboxylation of di(poly)ols (Eq. 9.11) [25]. The former route needs abundant and cheap hydrogen peroxide (market of *ca.* 800 kt/y, much lower than the market of carbonates). The direct oxidative carboxylation of olefins [26] (Eq. 9.12) avoids the synthesis of epoxides, but requires further development for avoiding the two-oxygen addition to the olefin that increases the complexity of the reactive system and causes loss of olefin.

$$R \underset{H}{\overset{}{\triangle}} H \quad + CO_2 \quad \xrightarrow{cat} \quad \text{(cyclic carbonate)} \tag{9.10}$$

$$HOCH_2CHCH_2OH \xrightarrow{CO_2} \text{(structure)} \tag{9.11}$$
$$\underset{OH}{|}$$

$$RCH_2=CH_2 + CO_2 + 1/_2 O_2 \longrightarrow \text{(structure)} \tag{9.12}$$

A way to win such drawbacks is the use of urea as a substitute of CO_2 in the synthesis of organic carbonates (Eq. 9.13): ammonia can be recovered and reused to make urea or employed in alternative reactions. Such reaction has no thermodynamic limitation and can be run to completion [27]. The catalytic carbonation of methanol to DMC with urea has been demonstrated at a demo scale [28].

$$2\,ROH\ (or\ HOCH_2CH_2OH) + H_2NCONH_2 \rightarrow (RO)_2CO\ (\ or\ \text{(structure)}\ ) + 2NH_3 \tag{9.13}$$

The carboxylation with urea has been developed to TRL around 7, while the reaction with CO_2 is still at TRL 3–5.

9.2.4.5 Carbamic Acid, RHN–CO₂H and Its Derivatives

$$RHN-C\overset{O}{\underset{OH}{\diagdown}} \qquad \underset{O}{\overset{RO}{\diagdown}}C-NHR \qquad RHN-C\overset{O}{\underset{OH}{\diagdown}} \qquad \underset{O}{\overset{HO}{\diagdown}}C-NHR$$

Carbamic acid   Carbamic esters   Carbamic acid dimers

Carbamic acid does not exist as a free species. Its dimers have been isolated (see text).

The **molecular ester derivatives** are stable and can be used for several applications as: fungicides, herbicides, pesticides in agriculture or in the cosmetic industry and other applications. Their market is difficult to size due to the myriad of products. The carbamic moiety $-NH-CO_2-$ is the base of **polyurethanes (PU)** which *global market* size was 55.2 BUS$ in 2018 and is projected to reach 87.6 BUS$ by 2026, exhibiting a CAGR of 6.0%. They find application as foams, packaging agents, thermal and sound insulators, components of cars, in materials for furniture, toys, *etc.*

Carbamic acid is a labile species that dissociates into the free amine (RNH_2) and CO_2, unless it is stabilized by dimerization [29]. Conversely, carbamic acid esters are stable entities used in several applications: cosmetics, agriculture, insecticides, monomers for polymers (polyurethanes).

In principle they can be produced by reacting amines (primary or secondary aliphatic amines) with CO_2 and an alkylating agent (Eq. 9.14) in presence of a base (the latter can be the amine itself).

$$RR'NH + R''X + CO_2 + B \rightarrow RR'N-CO_2R'' + BH^{+-}X \qquad (9.14)$$

Usually, polyurethanes are synthetized by reaction with diols $[HO(CH_2)_nOH, n > 2]$ of di-isocyanates, $OCN-R-NCO$, (see **DI9.1**), obtained by thermal or catalytic conversion of primary amine carbamates ($RHN-CO_2R' \rightarrow RNCO + R'OH$). Polymeric chains such as $HOR''O[(O)C-NH-R-NHC(O)OR''O]_n-$ are obtained. (see also D9.1)

As isocyanates are supposed to be toxic, alternative routes to such compounds are actively searched.

9.2.5   CCU and the Sustainable Production of Fuels

So far, in all reactions only an organic substrate and CO_2 have been used. Let us analyze now cases in which CO_2 is reacted with hydrogen.

9.2.5.1   Methanol, CH₃OH

$$
\begin{array}{c}
H \\
| \\
H-C-OH \\
| \\
H
\end{array}
$$

Has a worldwide market of the order of 85 Mt/y, [31] with several applications, from fuel to solvent, and as a feedstock for the synthesis of several chemicals including Cn hydrocarbons or unsaturated hydrocarbons.

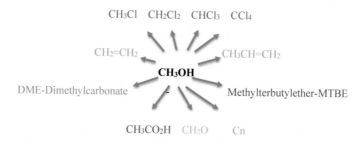

Fig. 9.7 Methanol as feedstock for the Chemical Industry. Chlorinated products are more and more out of use, ethene (left) and propene (right) are used in the polymer industry (PTE, PP) and other applications, DME finds several uses (see Organic Carbonates), MTBE is an additive to gasoline for improving the octane number, Acetic acid finds applications as reagent and in food industry, formaldehyde is the base for resins, Cn hydrocarbons are used as fuels. *Reprinted/adapted by permission from Ref. [6b] Copyright 2019*

Methanol, besides a potential use as fuel in thermal motors or in fuel cells, has a widespread use as chemical.

Figure 9.7 gives an idea of its multiple applications [30].

It is currently produced from Syngas (CO + 2H$_2$), but CO$_2$ could be used that requires one excess mol of hydrogen (Eq. 9.15).

Even if the conversion of CO$_2$ into methanol is exothermic by *ca.* 49 kcal mol^{-1}, a temperature higher than 513 K is necessary [32] in the synthesis of methanol (Eq. 9.15) for overcoming kinetic barriers, while more active and selective catalysts are necessary for improving the overall yield and selectivity of the process and reduce its environmental impact [33].

$$CO_2 + 3H_2 \rightarrow CH_3OH + H_2O \qquad (9.15)$$

The advantage of using CO$_2$ is that carbon is recycled instead of being sourced from fossil-C. Non-fossil-H$_2$ must be used. Methanol produced from CO$_2$ can be used for the synthesis of olefins using the "*methanol to olefin-MTO*" and "*methanol to propene-MTP*" processes [34]. In this way, large-scale polymers such as polyethene-PET (market of 150 Mt/y in 2017, [35]) and polypropene-PP (market of 73.7 Mt/y in 2017 [36]) could be entirely made from CO$_2$ with an utilization of some 880 Mt/y. In all cases [37], catalysts must be improved for increasing their selectivity, avoiding that a variety of olefins are obtained, from C2 to C8 with different abundance, and reducing the energetic and economic cost of separation. Even the electro-reduction of CO$_2$ in water to ethene (using Cu-electrodes [38]) has reached a conversion and selectivity level that may open to a potential exploitation.

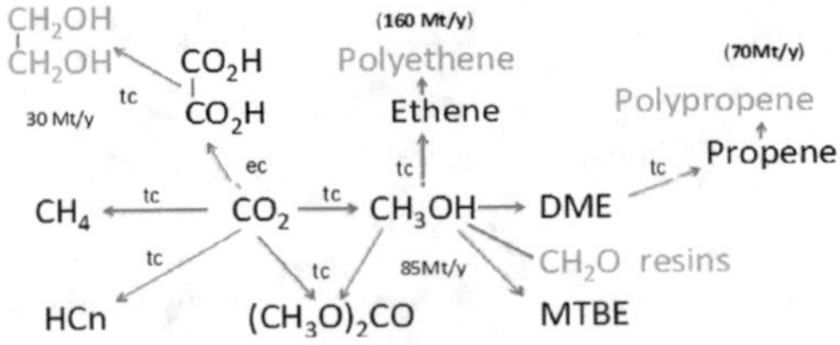

Scheme 9.2 A summary of CH_3OH-centered Value-Chains for CO_2 utilization. (*tc = thermal catalysis; ec = electro catalysis.* Figures in parentheses give the actual market of products made using fossil-C: in future they could be made from CO_2). *Reprinted/adapted by permission from Ref. [6b], Copyright 2019*

In Scheme 9.2 a representation of Value-Chains based on CO_2–CH_3OH is shown which gives a cumulative idea of the products that can be obtained from CO_2 using CH_3OH as intermediate.

The integrated approach can potentially convert over 1 Gt_{CO2}/y.

9.2.5.2 Dimethyl Ether, CH_3OCH_3, DME

DME is a low boiling liquid, used as an intermediate in the synthesis of methyl acetate, dimethyl acetate and light olefins or as a fuel for substituting diesel (40+% better cetane number). It is produced at a rate of *ca.* 6 Mt/y (2018). Most of it is produced in China.

DME is produced [39] upon methanol dehydration, (Eq. 9.16a) but the one-reactor technology based on CO_2 hydrogenation is very attractive (Eq. 9.16b) as reduces both CAPEX and OPEX [40].

$$2CH_3OH \rightarrow CH_3OCH_3 + H_2O \qquad (9.16a)$$

$$2CO_2 + 6H_2 \rightarrow CH_3OCH_3 + 3H_2O \, \Delta H^{\circ}_{298K} = -122.2 \, kJ \, mol^{-1} \qquad (9.16b)$$

In practice, the catalyst for CO_2 hydrogenation to methanol is coupled with the dehydration catalyst so to run both processes in a single reactor.

9.2.5.3 C–C Reduction Coupling

C–C reductive coupling of CO_2 to afford diols (ethene glycol, $HO–CH_2CH_2–OH$, EG) and poly-ols (glycerol and others) is possible by electrochemical reduction [41], even if at very low yield 0.50% (Toshiba). The interest in such process lays in the large market of EG (expected to reach 22 Mt/y in 2020 [42]) used as co-monomer in polymers such as polyesters and polyols. Its production from CO_2 would convert *ca.* 100 Mt/y of CO_2 into long-living materials. Electrochemically, even other C2 molecules, such as glyoxylic acid $–HO_2C–CHO$ (World Market (WM) 140 kt/y), glycolic acid $–HO_2C–CH_2OH$ (WM 100 kt/y), acetic acid $HO_2C–CH_3$ (WM 15 Mt/y) can be produced.

9.2.6 CCU in the Production of Large Scale Fuels

We shall now discuss the potential conversion of CO_2 into fuels, targeting three broad classes: CO, CH_4, liquid hydrocarbons. Each has its own peculiarities, as the reader will see.

9.2.6.1 Carbon Monoxide, CO

The reduction of CO_2 to CO requires high energy if carried out in the gas phase. As discussed in Chap. 2, the production of CO + "O" may occur when CO_2 is irradiated with a wavelength lower than 160 nm or is heated above 1000 °C [43]. Binding of CO_2 to a metal center greatly favor the deoxygenation of CO_2: the $\eta^2_{C,O}$-co-ordinated species is able to transfer one oxygen atom to an oxophile intra- (Eq. 9.17) or inter-molecularly (Eq. 9.18) at 30–60 °C [44].

$$LnM(CO_2) \rightarrow LnMO + CO$$
$$\rightarrow (L = \text{ligand such as phosphines, } PR_3; \ M = \text{Metal}) \quad (9.17)$$

$$LnM(CO_2) + Sub \rightarrow LnM(CO) + Sub$$
$$= O(Sub = \text{organic or inorganic substrate}) \quad (9.18)$$

As will be described in Chap. 10, solar energy can be concentrated to achieve high temperature used for the deoxygenation of CO_2 to CO, promoted by some metal oxides [45] (Eq. 9.19a, b). Coupling such reaction with water-splitting ($H_2 + 1/2O_2$), (Eq. 9.19c) Syngas (CO + H_2) is produced that can be used for the production of gasoline and diesel (Eq. 9.19d) [46]. Considering that both water vapor and CO_2 can be recovered from the atmosphere, the fuel produced is labeled "*Diesel from Air*".

$$MO_x + \text{Energy} \rightarrow MO_{x-1} + 1/2O_2 \quad (9.19a)$$

$$MO_{x-1} + CO_2 \rightarrow MO_x + CO \quad (9.19b)$$

$$MO_{x-1} + H_2O \rightarrow MO_x + H_2 \tag{9.19c}$$

$$xH_2 + yCO \rightarrow Liquid\ fuels \tag{9.19d}$$

CO has a large utilization in carbonylation reactions (introduction of the CO-moiety in organic substrates), in FT processes (see Chap. 2) and in the synthesis of methanol; its 2017 market was 2870 MUS$, with a forecast to grow at a 1.8% CAGR until 3310 MUS$ in 2025 [47].

9.2.6.2 Methane, CH$_4$

Methane is used today at a rate of 3850 Bm^3/y (2018) [48a].

It is used today mainly for the following applications: heating, home fires, source of heat in industries, production of Syngas for the chemical industry, other minor industrial uses. Major producers are USA and Russia.

Methane is extracted from natural wells as natural gas or is produced by fermentation of fresh vegetal waste. In both cases, methane must be separated from CO_2. Methane can be also produced through the hydrogenation of CO_2, (Eq. 9.20) a process known as the Sabatier-Senderens reaction [48].

The methanation of CO_2 is of great interest for both the automotive industry, as it represents a way to reduce the net emission of CO_2, and for biogas and natural gas valorization, possibly without preliminary separation of the two C1 molecules [49].

$$CO_2 + 4H_2 \rightarrow CH_4 + 2H_2O \quad \Delta H = -165\,kJ\,mol^{-1} \tag{9.20}$$

As shown in Eq. 9.20, the methanation of CO_2 is strongly exothermic and this rises the problem of controlling the heat released. New catalysts are needed for a better control [49] of the reaction conditions.

9.2.6.3 Cn Hydrocarbons and Higher Olefins

The production of Cn hydrocarbons by direct hydrogenation of CO_2 (Eq. 9.21) is very appealing

$$n\,CO + (2n+2)\,H_2 = H(CH_2)_nH + nH_2O \quad \Delta H = -166\,kJ\,mol^{-1} \tag{9.21}$$

for producing fuels for the transport sectors, including avio and navy, while recycling large quantities of C. Challenges are the difficulty to drive selectivity toward a single species. Catalysts based on iron-nanoparticle@carbon-nanotube (Fe@CNT) seem to have quite good potential [50] to drive the reaction under quasi C-neutral conditions even within the actual energy system frame. Indium-modified zeolites [51] give high yield (>78%) of Cn hydrocarbons with low methane selectivity. Cn *olefins* can be produced (54% light olefins with >50% conversion CO_2) adding K^+ as co-catalyst [50, 52].

Such reactions could be exploited even today if excess-hydrogen is used. Flared-H_2 amounts at *ca.* 8 Gm^3/y, as it represents an average 5.54% of the total flared gases (150–179 Gm^3) which cause the emission of over 450 Mt/yCO_2 [53]. Once renewable, low-cost hydrogen will be available, the conversion of CO_2 into Cn hydrocarbons or olefins may become an option economically and energetically viable, possibly clustering processes and operations for an efficient use of materials and even waste gases, with much benefit for the environment and our society.

9.2.7 Other Reactions

CO_2 can be used for the synthesis of a variety of compounds containing either the carboxylic $-CO_2-$ or the carbonylic $-CO-$ moiety. The use of CO_2 would allow to implement more direct syntheses, reducing waste production and even energy consumption, considered that the co-reagents are energetic molecules. Examples of the former reactions are the synthesis of lactones, pyrones or esters. Two examples (Eqs. 9.22a–9.22b) are reported below to show that even complex structures can be obtained with a single direct reaction [54]. In all such reactions, key issues are the lifetime of the catalyst and the selectivity toward a given product. High TRL (6–7) has been reached in the conversion of butadiene into the lactone [55] shown in Eq. 9.22a.

$$CH_2=CH-CH=CH_2 + CO_2 \xrightarrow{Pd\text{-cat}} \qquad\qquad\qquad\qquad (9.22a)$$

$$2\ R-\!\!\!\equiv\!\!\!-R\ +\ CO_2 \xrightarrow{cat} \qquad\qquad\qquad\qquad (9.22b)$$

Several other reactions of CO_2 with a variety of agents (Eqs. 9.23a–9.23d) can be found in the literature, but they are not yet at an application level and need much investigation for showing their viability in synthetic chemistry, reducing the complexity of the catalytic systems for saving energy and reducing waste and cost.

$$RNH_2 + CO_2 + H_2 \rightarrow RHNC\overset{O}{\underset{H}{\diagdown}} + H_2O \qquad (9.23a)$$

$$3R_3SiH + CO_2 \rightarrow CH_3OSiR_3 + R_3SiOSiR_3; \; CH_3OSiR_3 > > > CH_3OH$$
$$(9.23b)$$

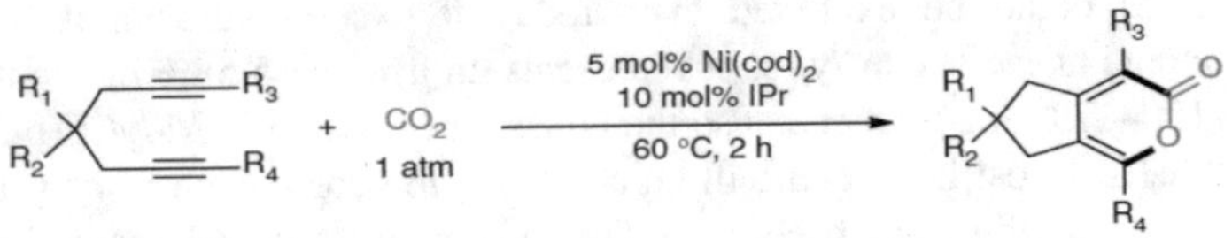

$$(9.23c)$$

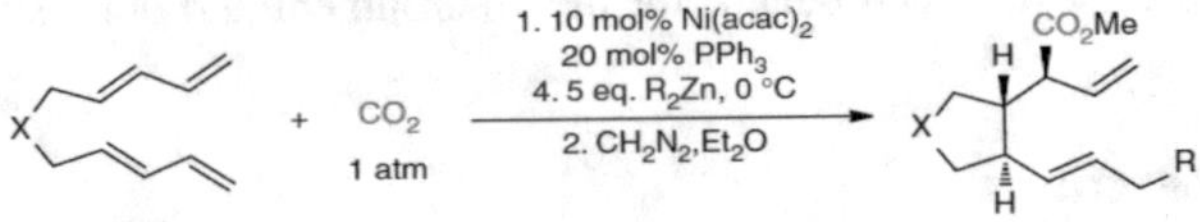

$$(9.23d)$$

In order such reactions, and many similar others, may reach the exploitation level, a number of key issues must find a solution: the life of the catalyst is often too short and their cost too high; expensive (energetically and economically) co-reagents are used which are lost after the first cycle; the yield and selectivity toward the target product still needs to be improved, the cost (energetic and economic) of isolation of the pure target product must be lowered. All above reactions are more suited for simplifying processes for the synthesis of high added value products with a niche market (a few kt/y) more than large scale ($\gg 1$ Mt/y) chemicals.

9.2.8 Electrochemical Reduction of CO_2

A route to convert CO_2 is its electrochemical reduction. In Electrochemistry, an oxidation and a reduction reaction are coupled. In the case of CO_2, it can be reduced to a variety of species which require a different energy (more negative is the potential, higher is the energy required for the process to occur). (Table 9.4).

The highest energy is required for the one-electron transfer to CO_2 to afford the CO_2^- species. This is due to the fact that the linear CO_2 molecule must be bent to the CO_2-radical anion, a process that requires energy (-0.4 eV). The reduction to the other species requires less energy because a different mechanism is active: the proton coupled to electron transfer-PCET ($H^+ + e^-$).

Table 9.4 Electrochemical reduction potential of CO_2 to several Products

	$E°$ [V] versus SHE at pH 7 in water
$CO_2 + e^- \rightarrow CO_2^-$	-1.9 (-2.1 in organic solvents)
$CO_2 + 2H^+ + 2e^- \rightarrow HCOOH$	-0.61
$CO_2 + 2H^+ + 2e^- \rightarrow CO + H_2O$	-0.52
$2CO_2 + 12H^+ + 12e^- \rightarrow C_2H_4 + 4H_2O$	-0.34
$CO_2 + 4H^+ + 4e^- \rightarrow HCHO + H_2O$	-0.51
$CO_2 + 6H^+ + 6e^- \rightarrow CH_3OH + H_2O$	-0.38
$CO_2 + 8H^+ + 8e^- \rightarrow CH_4 + 2H_2O$	-0.24
$2H^+ + 2e^- \rightarrow H_2$	-0.42

The process of interest for CO_2 conversion is the reduction of CO_2 coupled to oxidation of water to elemental oxygen (Eq. 9.24):

$$CO_2 + H_2O = \text{``}H_2CO\text{''} + O_2 \tag{9.24}$$

Equation 9.24 is the elemental step often used to represent the *natural photosynthesis*.

The electrochemical reduction of CO_2 can be coupled even to the conversion (oxidation) of organics. Electrochemical syntheses have been used for the synthesis of carboxylated species (we mention the following: carboxylation of halides to carboxylic acids, double carboxylation of olefins to dicarboxylic acids) or for the reduction to methanol and C2 (or even Cn) species with formation of C–C bonds (etheneglycol, $HOCH_2CH_2OH$; ethene, $CH_2=CH_2$). The two latter reactions have a great industrial interest.

9.3   CCU and Clustering of Processes

The production of fuels is the sector that will use large volumes of CO_2. The market of chemicals is some 15 times smaller. On the other hand, it must be emphasized that chemicals may have complex structures and processes on stream for their synthesis are multistep, energy-consuming and waste-producing: the use of CO_2 may reduce the Carbon Footprint-CF of a process by avoiding up to 2–3 times the amount of CO_2 fixed. This has been demonstrated by LCA studies [12] (see Appendix F). Moreover, a one-step process avoids organic waste with respect to a multi-step; reducing the production of organic waste means saving CO_2 emission as the fate of most of organic waste is burning often without real utilization of the heat produced. As we have already discussed, the conversion of CO_2 into energy products requires energy and hydrogen, both not originated from fossil-C. However, only if perennial energy sources are used to power the process and hydrogen is derived from water it makes sense to convert CO_2 into energy products for some selected sectors, such as avio-, navy-, heavy road-transport where electric motors (directly or indirectly) powered by solar energy cannot be conveniently used. Noteworthy, it would make sense to implement such reduction reactions even today if excess hydrogen is used. As already mentioned, flared-H_2 amounts at *ca.* 8 Gm^3/ y, as it represents an average 5.54% of the total flared gases (150–179 Gm^3) which cause the emission of over 450 Mt/yCO_2 [53]. However, the production of large-scale fuels from CO_2 requires non-fossil-H_2, such as low-cost PV-H_2 or H_2 that would be flared. Only under the latter conditions the conversion of CO_2 into Cn hydrocarbons or olefins may become an option economically and energetically viable.

The considerations just made, push us pay attention to an interesting issue: the opportunity of "*clustering processes and operations*" for an efficient use of materials and even waste gases, with much benefit for the environment and our society. Here, some key elements will be discussed for clarifying the frame in which such option can really contribute to recycling carbon and reducing CO_2 emission.

Most likely, the existing industrial organization will be revolutionized in future for generating better options of a cascade utilization of goods and residues. Industrial processes dispersed in separated sites rise the problem of transportation of specialty residues. In general, effluents and solid residues of an industrial process cannot be freely transported on road but can circulate within an industrial site. In a *circular economy frame*, effluents and residues can become "*secondary raw materials*" for another process. Clustering of processes and diverse activities will play a key role in order to optimize the utilization of raw materials and minimize waste production and CO_2 emission. We have always believed that *CO_2 is renewable carbon* [56], it can be recovered and cycled incessantly, as Nature has always done. Clustering of processes is a strategic approach to the efficient use of resources and valorization of "waste" streams. Approaching the conversion of CO_2 via "value-chains," more than single reactions, will give a new system perspective to CCU. Scheme 9.2 is an example of integration of processes for the production of chemicals and fuels. It shows how it is possible to connect processes for going from a putative waste (CO_2) to a variety of chemicals, materials, and fuels. If such processes are present on the same site a great advantage is produced in chain development in terms of economy of transport and storage.

Another example is given in Scheme 9.3 which shows how hydrogen can be managed in a process system: if processes are clustered, "residual" H_2 can be more easily cycled into a new process for CO_2 reduction opening to a range of opportunities. However, Scheme 9.3 shows that CO_2 (combined or not with oxygen) can be used as dehydrogenating agent (Scheme 9.3, left part, reactions 1–3) toward aliphatic hydrocarbons (CO_2 is a mild oxidant), namely in the coupling of methane or dehydrogenation of propane to propene or ethylbenzene to styrene, a process that is finding industrial exploitation with some demo-plants in Korea and China. Such process brings to the production of single C–C or double C=C bonds and hydrogen that could be used for the reduction of CO_2 to useful products in situ.

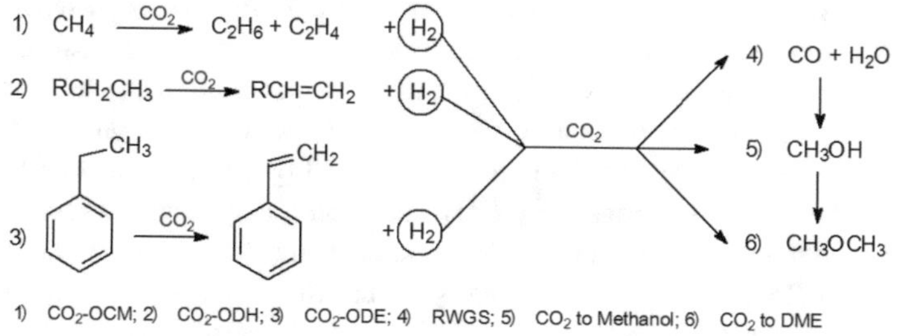

Scheme 9.3 Utilization of CO_2 as dehydrogenating agent and the use of produced H_2 for CO_2 reduction. *Reprinted from Ref. [44], Copyright (2016), with permission from Elsevier*

Such concept will be a driving force in future and process integration within the chemical industry and of the chemical industries with other sectors (utilization of biomass, biotechnology) will be the winning option (see Chap. 11). Carbon-circular economy and Bio-economy are strongly complementary and, if integrated, can multiply the positive effects, balancing some externalities [57]. On the other hand, CO_2 is at the basis of both and the integration can reinforce weak points of one with strong points of the other. Notably, Nature has not maximized energy efficiency (there is plenty of solar energy) toward a single product but toward the production of useful chemicals to support life; while the chemical industry has the attitude to maximize selectivity and energy efficiency (of fossil-C, a not infinite resource) toward a single product. Science and Technology will be able to merge these two powers for finding the solution to the needs of our society? We believe they can.

9.4   Conclusions

In this chapter, we have shown that our future lays on stepping away from fossil-C and developing an ability to use CO_2 and water, in other words developing "*An economy based on CO_2 and water*" [58]. Key questions are: at what extent and when will this be possible?

The use of CO_2 as building block for added value products, does not need hydrogen. Such products have short- (order of months, such as fuels), medium- (order of year, chemicals), long-(order of decades, organic polymers) life. The TRL level varies from 4 to 9 with the process considered, in all cases bottlenecks are well-known, as well as remedies, opening to a good perspective in the exploitation of discussed technologies in the short–medium term (5–10 years). Nevertheless, the market of such products is limited, ranging from actual 200 Mt/y CO_2 to over 350–400 Mt/y within 2030. Apparently, such use of CO_2 will not contribute significantly to controlling the CO_2 atmospheric level. Nevertheless, if we consider not the *used* but the *avoided* CO_2, assuming an average *avoided/used* ratio of 2.8 as discussed before, even considering the potential technology innovation, then we can conclude that close to *1 Gt/y* of *CO_2* will be *avoided by 2030*; this represents a more significant contribution which is coupled to a lower extraction of fossil-C.

In the longer term, let us say by 2040, a different perspective can be built [9, 58] on the fact that by then the availability of large volumes of PV-H_2 at a cost close to MR-H_2 may become a reality. The conversion of CO_2 into fuels may grow to high levels, (2–3.5 Gt/y). Coprocessing CO_2 and water may also serve to produce large-scale chemicals other than fuels. Of interest is the case discussed above of the synthesis of C2 and C3 olefins from CO_2 via electrolysis that would contribute to rise the amount of CO_2 used to *ca.* 1 Gt/y. A study carried out by the Catalyst Group [17] shows that combining the potential of all technologies it will be possible by 2040 avoid some 7–9 Gt_{CO2}/y; this is a very interesting target that will contribute to CC control.

And this is a ***Revolution based on CO_2***.

DI9.1 Synthesis of carbamates from CO_2

The development of synthetic methodologies that may afford carbamates from amines and CO_2 has an industrial interest as such process might supplant the process based on phosgene, a highly toxic substance, banned in several countries and whose transport on road is forbidden. Amine are very reactive with CO_2 as they are electron-rich (the nitrogen atom bears a non-bonding electron pair that may be used to attack the C-atom of CO_2). When CO_2 and primary or secondary aliphatic amines are contacted in an organic aprotic solvent, a white precipitate of ammonium carbamate is quickly formed (Eq. 9.25). The latter can, in principle, be alkylated to afford a carbamic ester and ammonium chloride (Eq. 9.26)

$$2\,RR'NH + CO_2 \rightarrow RR'N - CO_2^- NH_2RR'^+ \tag{9.25}$$

$$RR'N - CO_2^{-\,+}NH_2RR + R''X \rightarrow RR'N - CO_2R'' + NH_2RR'^{+\,-}X \tag{9.26}$$

$$RR'N - CO_2^{-\,+}NH_2RR' + R''X \rightarrow RR'NR'' + CO_2 + NH_2RR'^{+\,-}X \tag{9.27}$$

The bottleneck of such reaction is that the carbamate anion $RR'N-CO_2^-$ (Eq. 9.25) has two nucleophilic sites (Scheme 9.4); at the N atom that bears a non-bonding lone pair and at the negatively charged oxygen-atoms of the carboxylic moiety (the two oxygen atoms are equivalent).

The alkylation at the N-atom is a parasite reaction that produces an alkylated amine instead of the carbamic ester (Eq. 9.27). In the carbamate salt (either ammonium- or metal-carbamate) the cation can interact with the carboxylic group, reducing the ability of the anionic-O to react with the R''^+ cationic moiety of the alkylating agent $R''X$ (Scheme 9.4) [59].

Such parasite reaction can be prevented by using a suitable Crown-Ether (CE) (Scheme 9.5) that reduces the interaction of the metal cation (either Group 1 metal or ammonium) with the carbamate anion.

A good alternative to alkyl halides (that these days are not very popular and less and less used in the chemical industry) is the use of aliphatic carbonates (Eq. 9.28) that can be prepared from CO_2 and alcohols (see main text).

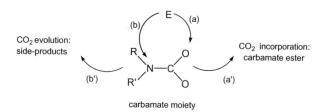

Scheme 9.4 Interaction of the ambiphilic carbamate moiety with an electrophile-E. *Reprinted from Ref. [59], Copyright (2016), with permission from Springer*

Scheme 9.5 How to avoid
the alkylation at the N-atom
and promote the O-alkylation.
*Reprinted from Ref. [59],
Copyright (2016), with
permission from Springer*

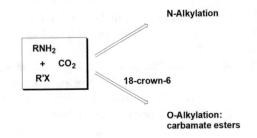

$2\ RNH_2 + CO_2 + 18\text{-crown-6} + R'X \longrightarrow RNHCO_2R' + [RNH_3(18\text{-crown-6})]X$ (a)

Mechanism:

$2\ RNH_2 + CO_2 + 18\text{-crown-6} \rightleftharpoons [RNH_3(18\text{-crown-6})]O_2CNHR$ (b)

$[RNH_3^-(18\text{-crown-6})]O_2CNHR + R'X \longrightarrow RNHCO_2R' + [RNH_3(18\text{-crown-6})]X$ (c)

$$RR'NH + (R''O)_2CO \rightarrow RR'N{-}CO_2R'' + R''OH \qquad (9.28)$$

Unfortunately, even such reaction can produce N-alkylation. Therefore, suitable promoters are necessary which promote the O-alkylation by preventing the N-alkylation. This is particularly necessary in industrial processes for the synthesis of carbamates and isocyanates. Alkylation of amines cause not only loss of the substrate, but also the formation of more complex reaction mixtures that need more energy for the separation of target products.

Di-carbamates of primary amines find an industrial application as source of di-isocyanates (Scheme 9.6).

The carbamation of aromatic amines requires active and selective catalysts. Bio-mimetic mixed anhydrides [60–63], structurally analogous to the enzyme carboxyphosphate (Scheme 9.7) active in the elimination of ammonia in living organisms, are effective and selective catalysts in such reaction.

Isocyanates are used as co-monomers with diols [$HOCH_2(CH_2)nCH_2OH$, n = 0, 1, 2…] for the production of polyurethanes that have a market of several Mt/y and an estimated 79 BUS$ value at 2021 (53 BUS$ in 2014), used in a variety of industrial sectors from automotive to agriculture, personal care, fillers, adhesives,

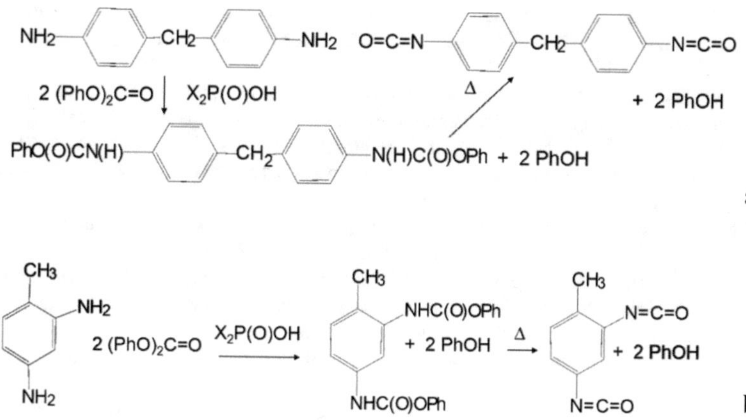

Scheme 9.6 Conversion of aromatic diamines into isocyanates via carbamation

Scheme 9.7 The enzyme Carboxyphosphate (a) and its mimetic mixed-anhydride (b). *Reprinted/adapted by permission from Ref. [1b], Copyright 2019*

sealants etc. [64]. These days, the tendency is to shift from fossil-C-based monomers to bio-sourced molecules, using polyols and amines derived from biomass. Such move has a key role for the sustainability of the chemical and polymer industry and opens the way to an extensive carbon recycling, including the exploitation of CCU.

References

1. https://www.CO2.earth/
2. (a) Aresta M, Galatola M (2001) Life cycle analysis applied to the assessment of the environmental impact of alternative synthetic processes. J Cleaner Prod 7:181–193; (b) Aresta M, Caroppo A, Dibenedetto A, Narracci M (2002) Life Cycle Assessment (LCA) applied to the synthesis of methanol. Comparison of the use of syngas with the use of CO_2 and dihydrogen produced from renewables. In: Maroto-Valer M (ed) Envrironmental challenges and greenhouse gas control for fossil fuel utilization in the 21st century. Kluwer Academic/Plenum Publishers, New York; (c) Artz J, Müller TE, Thenert K, Kleinekorte J, Meys R, Sternberg A, Bardow A, Leitner W (2018) Sustainable conversion of carbon dioxide: an integrated review of catalysis and life cycle assessment. Chem Rev 118(2),:434–504

3. Aresta M, Quaranta E, Tommasi I, Giannoccaro P, Ciccarese A (1995) Enzymatic versus chemical carbon dioxide utilization. Part I. The role of metal centres in carboxylation reactions. Gazz Chim Ital 125(11):509–538

4. Helmenstine AM (2018) Heat of formation or standard enthalpy of formation table, thought co. https://www.thoughtco.com/common-compound-heat-of-formation-table-609253

5. (a) Rotman D (2015) MIT Technol Rev https://www.technologyreview.com/s/425489/a-future-of-fossil-fuels/; (b) Abas N, Kalair A, Khan H (2015) Review of fossil fuels and future energy technologies. Futures 69:31–49

6. (a) https://knoema.com/infographics/smsfgud/bp-world-reserves-of-fossil-fuels; (b) Aresta M, Nocito F (2019) Large scale utilization of carbon dioxide: from its reaction with energy rich chemicals to (Co)-processng with water to afford energy rich products. Opportunities and barrers. In: Aresta M, Kawi S, Karimi IA (eds) An economy based on carbon dioxide and water. Springer Publ. ISBN 978-3-030-15868-2, chapter 1

7. https://www.power-technology.com/comment/global-pv-capacity-expected-reach-969gw-2025/

8. (a) Goto Y, Wang Q (2018) A particulate photocatalyst water-splitting panel for large-scale solar hydrogen generation. Joule 2(3):509–520; (b) Pinaud BA, Benck JD, Seitz LC, Forman AJ, Chen Z, Deutsch TG, James BD, Baum KN, Baum GN, Ardo S (2013) Technical and economic feasibility of centralized facilities for solar hydrogen production via photocatalysis and photoelectrochemistry. Energy Environ Sci 6(7):1983–2002

9. Aresta M, Dibenedetto A, Angelini A (2013) The changing paradigm in CO_2 utilization. J CO_2 Util 3:65–73

10. Aresta M, Quaranta E (1997) Carbon dioxide: a substitute for phosgene. ChemTech 27(3):32–40

11. https://www.uptodate.com/contents/carbon-monoxide-poisoning

12. Rupesh S, Muraleedharan C, Arun P (2016) Exergy and energy analyses of Syngas production from different biomasses through air-steaming gasification. Front Energy 1–13

13. https://www.marketresearchfuture.com/reports/formic-acid-market-1132

14. Tanaka R, Yamashita M, Nozaki K (2009) Catalytic hydrogenation of carbon dioxide using Ir (III)−pincer complexes. J Am Chem Soc 131(40):14168–9

15. Wesselbaum S, Hintermaier U, Leitner W (2012) Continuous-flow hydrogenation of carbon dioxide to pure formic acid using an integrated $scCO_2$ process with immobilized catalyst and base. Angew Chem Int Ed 51:8585–8588

16. Moret S, Dyson P, Laurenczy G (2014) Direct synthesis of formic acid from carbon dioxide by hydrogenation in acidic media. Nat Commun 5:1–7

17. https://www.mordorintelligence.com/industry-reports/global-market-for-surfactants-industry

18. Liang Y-F, Steinbock R, Yang L, Ackermann L (2018) Continuous visible light-photo-flow approach for manganese-catalyzed (Het)Arene C−H Arylation. Angew Chem Int 57:10625–10629

19. (a) https://www.alliedmarketresearch.com/acrylic-acid-market; (b) Alvarez R, Carmona E, Galindo A, Gutierrez E, Marin JM, Monge A, Poveda ML, Ruiz C, Savariault JM (1989) Formation of carboxylate complexes from the reactions of CO_2 with ethylene complexes of molybdenum and tungsten. X-ray and neutron diffraction studies. Organometallics 8 (10):2430–2439; (c) Aresta M, Pastore C, Giannoccaro P, Kovacs G, Dibenedetto A, Papai I (2007) Evidence for spontaneous release of acrylates from a transition-metal complex upon coupling ethene or propene with a carboxylic moiety or CO_2. Chem Eur J 13(32):9028–9034; (d) Lejkowski ML, Lindner R, Kageyama T, Bodizs GE, Plessow PN, Mueller IM, Schaefer A, Rominger F, Hofmann P, Futter C, Schunck SA, Limbach M (2012) The first catalytic synthesis of an acrylate from CO_2 and an alkene–a rational approach. Chem Eur J 18 (44):14017–14025; (e) Wang X, Wang H, Sun Y (2017) Synthesis of acrylic acid derivatives from co_2 and ethylene. Chem 3:11–228; (f) Li Y, Liu Z, Cheng R, Liu B (2018) Mechanistic aspects of acrylic acid formation from CO_2–ethylene coupling over palladium—and nickel—based catalysts. ChemCatChem 10(6):1420–1430

20. Aresta M, Dibenedetto A, Quaranta E (2016) See for example chapter 6 in reaction mechanisms for carbon dioxide conversion. Springer
21. Aresta M, Dibenedetto A, Angelini A, Papai I (2015) Reaction mechanisms in the direct carboxylation of alcohols for the synthesis of acyclic carbonates. Top Catal 58(1):2–14
22. Aresta M, Dibenedetto A, Dutta A (2017) Energy issues in the utilization of CO_2 in the synthesis of chemicals: the case of the direct carboxylation of alcohols to dialkyl-carbonates. Cat Today 281:345–351
23. Dibenedetto A, Aresta M, Angelini A, Etiraj J, Aresta BM (2012) Synthesis characterization and use of NbV/CeIV-mixed oxides in the direct carboxylation of ethanol by using pervaporation membranes for water removal. Chem-A Eur J 18(33):10524–10534
24. Della Monica F, Buonerba A, Capacchione C (2019) Homogeneous iron catalysts in the reaction of epoxides with carbon dioxide. Adv Synth Catal 361:265–282
25. Aresta M, Dibenedetto A, Nocito F, Pastore C (2006) A study on the carboxylation of glycerol to glycerol carbonate with carbon dioxide: the role of the catalyst, solvent and reaction conditions. J Mol Cat 257(1–2):149–153
26. (a) Aresta M, Quaranta E, Ciccarese A (1987) Direct synthesis of 1,3-benzodioxol-2-one from styrene, dioxygen and carbon dioxide promoted by Rh(I). J Mol Cat 41:355–359; (b) Dibenedetto A, Aresta M, Distaso M, Pastore C, Venezia AM, Liu CJ, Zhang M (2008) High throughput experiment approach to the oxidation of propene to propene oxide with transition metal oxides as O-donors. Catal Today 137:44–51; (c) Gabriele B, Mancuso R, Salerno G, Ruffolo G, Costa M, Dibenedetto A (2009) A novel and efficient method for the catalytic direct oxidative carbonylation of 1, 2-and 1, 3-diols to 5-membered and 6-membered cyclic carbonates. Tetrahedron Lett 50(52):7330–7332
27. Angelini A, Dibenedetto A, Curulla-Ferre D, Aresta M (2015) Synthesis of diethylcarbonate by ethanolysis of urea catalysed by heterogeneous mixed oxides. RSC Adv 5(107):88401–88408
28. Wang M, Wang H, Zhao N, Sun Y (2007) High-yield synthesis of dimethyl carbonate from urea and methanol using a catalytic distillation process. Ind Eng Chem Res 46(9):2683–2687
29. Aresta M, Ballivet-Tkatchenko D, Belli-Dell'Amico D, Bonnet MC, Boschi D, Calderazzo F, Faure R, Labella L, Marchetti F (2000) Isolation and structural determination of two derivatives of the elusive carbamic acid. RSC Chem Commun 13:1099–1100
30. Tian P, Wei Y, Ye M, Liu Z (2015) Methanol to Olefins (MTO): from fundamentals to commercialization. ACS Catal 5(3):1922–1938
31. https://www.technology.matthey.com/article/61/3/172-182/
32. Ma J, Sun NN, Zhang XL, Zhao N, Mao FK, Wei W, Sun YH (2009) A short review of catalysis for CO_2 conversion. Catal Today 148:221–223
33. Olah GA (2013) Towards oil independence through renewable methanol chemistry. Angew Chem Int Ed 52(1):104–107
34. https://www.engineering-airliquide.com/methanol-and-derivatives
35. http://blogs.platts.com/2017/09/07/infographic-whats-store-global-polyethylene-polypropylene-2027/
36. http://www.chemengonline.com/methanol-to-propylene-technology/
37. Inui T, Phatanasri S, Matsuda H (1990) Highly selective synthesis of ethene from methanol on a novel nickel-silicoaluminophosphate catalyst. JCS Chem Comm 3:205–206
38. Peng Y, Wu T, Sun L, Nsanzimana JMV, Fisher AC, Wang X (2017) Selective electrochemical reduction of CO_2 to ethylene on nanopores-modified copper electrodes in aqueous solution. ACS Appl Mater Interfaces 9(38):32782–32789
39. http://www.methanol.org/wp-content/uploads/2016/06/DME-An-Emerging-Global-Fuel-FS.pd
40. Frusteri F, Cordaro M, Cannilla C, Bonura G (2015) Multifunctionality of Cu–ZnO–ZrO$_2$/H-ZSM5 catalysts for the one-step CO_2-to-DME hydrogenation reaction. Appl Catal B Env 162:57–65

41. Tamura J, Ono A, Sugano Y, Huang C, Nishizawa H, Mikoshiba S (2015) Electrochemical reduction of CO_2 to ethylene glycol on imidazolium ion-terminated self-assembly monolayer-modified Au electrodes in an aqueous solution. Phys Chem Chem Phys 17 (39):26072–26078

42. http://www.grandviewresearch.com/industry-analysis/ethylene-glycols-industry

43. Marxer D, Furler P, Tacacs M, Steinfeld A (2017) Solar thermochemical splitting of CO_2 into separate streams of CO and O_2 with high selectivity, stability, conversion, and efficiency. Energy Environ Sci 10(5):1142–1149

44. Aresta M, Dibenedetto A, Quaranta E (2016) State of the art and perspectives in catalytic processes for CO_2 conversion into chemicals and fuels: the distinctive contribution of chemical catalysis and biotechnology. J Catal 343:2–45

45. (a) Bork AH, Kubicek M, Struzik M, Rupp JLM (2015) Perovskite $La_{0.6}Sr_{0.4}Cr_{1-x}Co_xO_{3-\delta}$ solid solutions for solar-thermochemical fuel production: strategies to lower the operation temperature. J Mater Chem 3(30):15546–15557; (b) Rao CNR, Dey S (2015) Generation of H_2 and CO by solar thermochemical splitting of H_2O and CO_2 by employing metal oxides. J Solid State Chem 242(2):107–115

46. Mostrou S, Buchel R, Pratsinis SE, van Bokhoven JA (2017) Improving the ceria-mediated water and carbon dioxide splitting through the addition of chromium. Appl Catal A: General 537:40–49

47. (a) https://www.mordorintelligence.com/industry-reports/global-synthesis-gas-Syngas-market-industry; (b) https://www.reportlinker.com/syngas-reports; (c) https://www.marketwatch.com/press-release

48. (a) https://www.statista.com; (b) Sabatier P, Senderens JB (1902) New synthesis of methane. J Chem Soc 82:333

49. (a) Jürgensen L, Ehimen EA, Born J, Holm-Nielsen JB (2015) Dynamic biogas upgrading based on the Sabatier process: thermodynamic and dynamic process simulation. Bioresource Technol 178:323–329; (b) Stangeland K, Kalai D, Li H, Yu Z (2017) CO_2 methanation: the effect of catalysts and reaction conditions. Energy Procedia 105:2022–2027; (c) Brooks KP, Hu J, Zhu H, Kee RJ (2007) Methanation of carbon dioxide by hydrogen reduction using the sabatier process in microchannel reactors. Chem Eng Sci 62:1161–1170; (d) Kirchner J, Katharina J, Henry A, Kureti LS (2018) Methanation of CO_2 on iron based catalysts. Appl Catal B Env 223:47–59; (e) Visconti CG (2010) Reactor for exothermic or endothermic catalytic reaction WO2010/130399; (f) Visconti CG (2014) Multi-structured reactor made of monolithic adjacent thermoconductive bodies for chemical processes with a high heat exchange WO2014/102350

50. Mattia D, Jones MD, O'Byrne JP, Griffiths OG, Owen RE, Sackville E, McManus M, Plucinski P (2015) Towards carbon-neutral CO2 conversion to hydrocarbons. Chemsuschem 8(23):4064–4072

51. Gao P, Li S, Bu X, Sun Y (2017) Direct conversion of CO_2 into liquid fuels with high selectivity over a bifunctional catalyst. Nat Chem 9(10):1019–1024

52. (a) Satthawong R, Koizumi N, Song C, Prasassarakich P (2013) Bimetallic Fe–Co catalysts for CO_2 hydrogenation to higher hydrocarbons. J CO_2 Util 3–4:102–106; (b) Visconti CG, Martinelli M, Falbo L, Infantes-Molina A, Lietti L, Forzatti P, Iaquaniello G, Palo E, Picutti B, Brignoli F (2017) CO_2 hydrogenation to lower olefins on a high surface area K-promoted bulk Fe-catalyst. Appl Catal B 200:530–542

53. Emam EA (2015) Gas flaring in industry: an overview. Pet Coal 57(5):532–555

54. Aresta M, Dibenedetto A, Quaranta E (2016) Reaction mechanisms in carbon dioxide conversion, chapter 5. Springer

55. (a) Pitter S, Dinjus E (2011) Phosphinialkyl nitriles as hemilabile ligands: new aspects in the homogeneous catalytic coupling of CO_2 and 1,3-butadiene. J Mol Catal A Chem 125:39–45; (b) Behr A, Henze H (2011) Use of carbon dioxide in chemical syntheses via a lactone intermediate. Green Chem 13:25–39

56. Aresta M, Forti G eds (1987) Carbon dioxide as a source of carbon: biochemical and chemical uses. Nato Sci Ser C

57. Dibenedetto A, Stufano P, Nocito F, Aresta M (2011) RuII-mediated hydrogen transfer from aqueous glycerol to co2: from waste to value-added products. Chemsuschem 4(9):1311–1315

58. Aresta M, Karimi I, Kawi S (2019) An economy based on CO_2 and Water. Springer

59. Aresta M, Dibenedetto A, Quaranta E (2016) Reaction mechanisms in carbon dioxide conversion, chapter 3. Springer

60. Aresta M, Dibenedetto A, Quaranta E (1998) Reaction of aromatic diamines with diphenylcarbonate catalyzed by phosphorous acids: a new clean synthetic route to mono- and dicarbamates. Tetrahedron 54(46):14145–14156

61. Aresta M, Bosetti A, Quaranta E (1996) Procedimento per la produzione di carbammati aromatici. Ital Pat Appl 002202

62. (a) Aresta M, Dibenedetto A, Quaranta E (2001) Selective carbomethoxylation of aromatic diamines: with mixed carbonic acid diesters in the presence of phosphorous acids. Green Chem 1(5):237–242; (b) Aresta M, Dibenedetto A, Quaranta E, Boscolo M, Larsson R (2001) The kinetics and mechanism of the reaction between carbon dioxide and a series of amines: observation and interpretation of an isokinetic effect. J Mol Catal A Chem 174(1–2):7–13

63. (a) Aresta M, Dibenedetto A (2002) Mixed anhydrides: key intermediates in carbamates forming processes of industrial interest. Chem A Eur J 8(3):685–690; (b) Aresta M, Dibenedetto A (2002) Development of environmentally friendly sytnthese: use of enzymes and biomimetic systems for the direct carboxylation of organic substrates. Rev Mol Biotechnol 90:113–128

64. https://www.grandviewresearch.com/industry-analysis/polyurethane-pu-market

Solar Chemistry and CO$_2$ Conversion

10

Abstract

The utilization of solar energy for driving reactions in which CO$_2$ is converted into energy-rich products is discussed in this chapter. Three approaches are considered: i. electrolysis of water by photovoltaic energy to generate hydrogen used, in turn, for the hydrogenation of CO$_2$ to gaseous or liquid fuels; ii. use of concentrated solar power for splitting CO$_2$ and H$_2$O and making Syngas; iii. the direct co-processing of water and CO$_2$ to energy-rich products under solar irradiation. The former option can be applied in the short term as both the electrolysis and CO$_2$ conversion are well-known processes; the second needs some further improvement of both the technology for concentrating solar energy and catalysts for H$_2$O and CO$_2$ splitting and may be applied in the medium term; the latter will most likely be exploited in the medium–long term as new, effective and stable photocatalysts must be discovered and advanced reactors-concentrators of solar light need to be developed.

10.1 Introduction

Over one century ago (1912) Giacomo Ciamician, an Italian Professor at the University of Bologna (Italy), in the *incipit of his paper* "*La Fotochimica dell'Avvenire*" (Photochemistry of Future) [1] wrote: "*The modern civilization is daughter of fossil carbon; the latter gives to humanity the solar energy in its most concentrated form; accumulated during centuries the modern man uses it with much avidity and careless lavishness for taking control of the world.*" He concluded by saying "*omissis…. I believe that the industry should use all forms of energy that Nature makes available; so far the modern civilization has mainly used fossil energy: maybe it should be better to use the other forms of available energies?*".

© Springer Nature Switzerland AG 2021
M. Aresta and A. Dibenedetto, *The Carbon Dioxide Revolution*,
https://doi.org/10.1007/978-3-030-59061-1_10

After more than one century, the Ciamician words are very actual. The urgent need to shift from a *linear economy* to a *circular economy* requires moving from fossil energy to perennial energies such as solar-, wind-, geothermal- and hydro-power or to renewable energy (biomass) for powering our life.

The paper of Ciamician was very future looking, in fact, it says: *"omissis....In desert areas where climatic conditions and the low quality of soil prevent any vegetal culture, artificial photochemistry will add value. The arid areas will be populated by industries without smog and chimneys: glass tubes and greenhouses - rooms made of glass- will catch solar radiations and in such transparent reactors photochemical processes will occur that were distinctive of plants and that humans will have learnt to use: industry will be able to fasten such processes because Nature is not in a hurry, but humanity is."*

However, the concept of *solar chemistry* is old of more than 100 years and it is joined to the vision of an *enhanced photosynthesis*! Today we are in the urgent need of implementing such concepts.

Solar chemistry is a vision and at the same time a necessity for shifting to a *circular economy*. Solar energy can power industrial processes in many different ways, e.g., thermal, photochemical, electrochemical, bioelectrochemical, photo-bioelectrochemical, each requiring peculiar operative conditions and producing different classes of products.

10.2   Utilization of Solar Energy for Driving Chemical Reactions

The large-scale storage of solar energy into chemical bonds, mimicking Nature, despite old and recent efforts, is still today a visionary concept. Nevertheless, solar-driven processes might find a large application that would save fossil resources and avoid the formation of huge amounts of CO_2. EUCheMS, the European Association of Chemical and Molecular Sciences, and DFG-Germany have recently published a white paper [2] highlighting the fields where solar-driven chemistry would be more effective and the relevant benefits. The conversion of CO_2 into chemicals, materials and fuels powered by solar radiations is one of the elective applications. Chapter 5 has described that solar energy can be used for generating either high-temperature heat or electricity: both can be applied to CO_2 conversion. In addition, solar radiations can be directly used for driving photochemical processes. However, solar energy can be used in three main different ways for driving the conversion of CO_2:

- Photochemical reactions (direct use of solar radiations)
- PV-driven processes, divided into three subsets

  o Electrochemical
  o Photoelectrochemical

   o Electrochemical and Photoelectrochemical coupled with biosystems (see Chap. 11)

- High-temperature processes driven by Solar Power Concentrators.

Each of these three technologies will be discussed and examples of application are presented.

10.3   The Solar Spectrum

Figure 10.1 shows the solar spectrum. From left to the right, Ultraviolet (UV), Visible (Vis), and Infrared (IR) regions are shown with their characteristic wavelengths and abundance. Clearly, the abundance depends on whether we measure the spectrum at sea level or outside the atmosphere. UV represents only *ca.* 8% of the total emitted radiations, while Vis (42.3%) and IR (49.4%) represent the largest part of it. UV radiations are used in chemistry: for example, TiO_2 (Titanium dioxide) is largely used as photocatalyst for the oxidation of organic toxic compounds under UV-irradiation. As shown in Fig. 10.1, UV represents only a tiny part of the solar radiation; therefore, for pushing the efficiency of the process, catalysts should be used as concentrators of the radiations. Most simply, UV lamps are used instead of using direct solar light in industrial applications, leaving to solar light the duty when very large surface applications are targeted, such as in the autocleaning of façades of buildings [3]. Visible radiations are used in photosynthetic pathways in Nature and are wished to drive Solar Driven Processes (SDP). Developing photocatalysts which operate under Vis-light is a key target for developing useful SDPs in the chemical industry. IR radiations are used for driving thermal reactions more than photochemical processes.

   Due to the low density of solar energy, in order to increase the number of photons that hit the target, devices which focus the solar radiations in a spot are often used. Alternatively, Xe-lamps are used as solar simulators, eventually coupled to filters for a better selection of wavelengths. Such lamps are fed with electricity (if PV is used, the source of energy will be the Sun) and emit radiations of wavelength similar to those of the solar spectrum [4]. In this case, the light intensity can be higher than that of the Sun. Using filters, it is possible to cut UV or IR radiations or even narrow the Vis-light to some of its components. Visible light is also known as "*white light*", as it has no color due to the combination of the radiations from blue to red (the colors of the "*rainbow*" that can be clearly seen in Fig. 10.1 in the Vis portion of the spectrum).

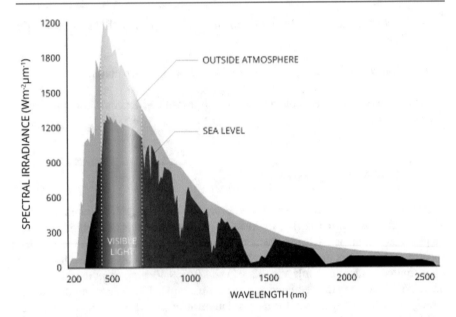

Fig. 10.1 Solar Spectrum. From the left: UV (200–400 nm), Vis (400–700 nm) IR (700–2500 nm) [3b]

10.4   Photochemical Reactions for CO$_2$ Conversion

For convenience of the reader, we report here again the table of the reduction potentials of CO$_2$ in water (Table 10.1) as they are relevant to the use of solar light for driving chemical reactions. The solar light-driven reduction of CO$_2$ has been attempted using solid and soluble catalysts and is a complex process (Fig. 10.2). In solution, it was described by Nobel Laureate J.M. Lehn in 1983 using [Re(dipy)(CO)$_3$Cl] (Fig. 10.3) and Ru-metal systems [5].

Table 10.1 Potential of reduction of CO$_2$ in water

Reaction	$E°$ [V] versus SHE *at pH 7 in water*
1. $CO_2 + e^- \rightarrow CO_2^-$	−1.9 (−2.1 in organic solvents)
2. $CO_2 + 2H^+ + 2e^- \rightarrow HCOOH$	−0.61
3. $CO_2 + 2H^+ + 2e^- \rightarrow CO + H_2O$	−0.52
4. $2CO_2 + 12H^+ + 12e^- \rightarrow C_2H_4 + 4H_2O$	−0.34
5. $CO_2 + 4H^+ + 4e^- \rightarrow HCHO + H2O$	−0.51
6. $CO_2 + 6H^+ + 6e^- \rightarrow CH_3OH + H2O$	−0.38
7. $CO_2 + 8H^+ + 8e^- \rightarrow CH_4 + 2H_2O$	−0.24
8. $2H^+ + 2e^- \rightarrow H_2$	−0.42

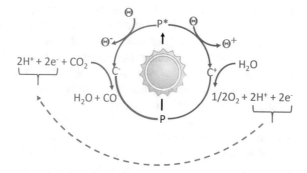

Fig. 10.2 Photochemical reduction of CO_2 in water in presence of a photocatalyst (P = photosensitizer; Θ = charge transfer system; C^+ = oxidation catalytic center; C^- = Reduction catalytic center)

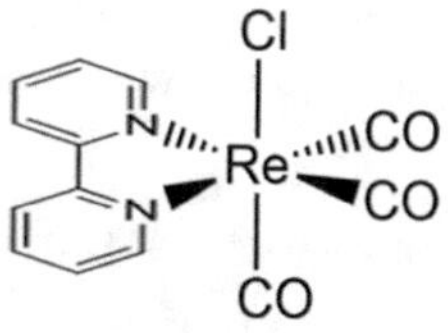

Fig. 10.3 The re-complex used by J.M. Lehn for CO_2 photo-reduction

It is evident from Table 10.1 that unless the potential is finely tuned, reactions "2, 3, 5, 8" can occur simultaneously, because they are very close in energy. This was verified by Lehn and others who have shown that CO (Entry 3), formates (Entry 2) and H_2 (Entry 8) are often coproduced [6], rising some selectivity issues. It is worth to note that CO and H_2 are not very soluble in water, while formates are. However, if CO and H_2 could be produced in such a way to approximate the Syngas composition, this would be an interesting way to produce Syngas and from it methanol or liquid fuels, but it is not so, the composition varies in a random mode with the reaction conditions. The CO_2 photoreduction to CO is only apparently a simple reaction; it results in a combination of several steps, actors and functions as shown in Fig. 10.2.

The first step in the cycle is the uptake and concentration of solar energy (photons) which is made by photon adsorbers/antennae. Such energy is used for the excitation of the Photosensitizer center (P) coupled to the catalytic centers. P* (the excited photosensitizer) causes a charge separation (e^- and $hole^+$), the electron moves to the catalytic center-C^- and is used for CO_2 reduction, while the $hole^+$ causes the water oxidation at-C^+ to afford O_2 and "$H^+ + e^-$", which are used in CO_2 reduction. The photosensitizer can even be a metal center. The pyridine-aromatic rings linked to the metal center (Fig. 10.3) may work as photon capture agents and eventually as concentrators. A variety of systems exist which operate in homogeneous conditions with the catalyst and CO_2 in the same (most likely liquid) phase; a lot of work is going on for developing more efficient metal-complexes which potential can be finely tuned for matching the potential of one of the reduction reactions listed in Table 10.1, so to develop selective reduction processes.

In addition to homogeneous systems, heterogeneous systems (two-three phase systems: gas, solution, solid) are effective which are made of "*semiconductors*" (binary metal oxides, ternary or multiple metal oxides, metal salts, metal oxides on graphene or graphene oxide, etc.) which are able to uptake light and produce an "*exciton*" (e^- + h^+) that drives the CO_2 reduction in presence of a (e^- + H^+) *donor*. Often organic molecules (which cost may be higher than the value of reduced forms of CO_2) are oxidized as model sacrificial agents, but water should be used in the real world. Figure 10.4 gives an example of how a semiconductor works. It is worth to note that the formed "*exciton*" can undergo recombination of the "e^- − h^{+}" couple (right part of the square) and the radiation energy up-taken in the excitation (left part of the square) is released as heat. Such recombination nullifies the excitation. The stabilization of the charge separation is one of the hot themes in this field. Often this is obtained by combining two different materials which can avoid the collapse of the exciton.

Insulators, conductors and semiconductors.

The difference among conductors, semiconductors and insulators is in the energy gap between the valence band (occupied energy levels) and the conduction band (empty energy levels). In conductors, such bands are very close in energy or even overlap, so that electrons can pass easily from levels occupied in the ground state to higher energy levels. Metals such as Cu, Al, Ag are example of excellent conductors. It is well known that Cu wires are used for electricity transmission. Silver is even a better conductor than copper. In semiconductors, the gap is such that a reasonable energy input (irradiation with VIS or UV light) is enough for electrons can flow from the ground state to excited levels with formation of a exciton "e⁻ + h⁺". Oxides, sulphides other compounds behave as such. In insulators (such as diamond), the energy gap is high enough to prevent the jump of electrons from the valence to the conduction energy levels.

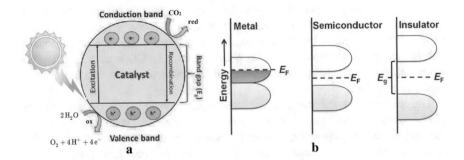

Fig. 10.4 How a semiconductor (solid photocatalyst) works (**a**). Comparison of conductors, semiconductors, and insulators (**b**). E_F = Fermi energy; E_G = Energy gap. The green area indicates the valence band, the white the conduction band

Semiconductors have found several applications in CO$_2$ conversion, either in carboxylation reactions or in reduction reactions. A number of selected cases will be discussed, to present the variety of applications and photocatalysts. In photocatalysis, an important point is the *efficiency in light utilization*. It has been anticipated that plants have a low efficiency in the use of solar light: only around 1.8–2.2% of the total energy that impacts the leaf is converted into chemical energy. Microalgae are better users: the photo-efficiency may reach 8% in natural environments and even 12% in photobioreactors (see Chap. 11). Man-made systems have been for long time confined in the range 0.1–0.5%, much below plants. The target is to do better than Nature, by reaching an efficiency threshold of 15–20%. This is already done with PV-systems in the *solar energy conversion to electricity (StE)*, that are now targeting 40% (see Chap. 5). The objective is to do the same within the direct conversion of solar energy to reduce carbon dioxide to chemicals (StC) or fuels (StF).

Two classes of reactions will be presented: the direct photocarboxylation of organic substrates with CO$_2$ and the photoreduction of CO$_2$ to higher energy products.

The photocarboxylation of acetylacetone, CH$_3$COCH$_2$COCH$_3$-acac, with visible light in presence of Graphene Oxide-GO (Fig. 10.5) loaded with CuO, is a reaction that implies the CH activation ($\Delta G^\circ_{C-H} = 414$ kJ mol^{-1}). It is worth to note that both the -CH$_2$- and -CH$_3$ moieties, contain hydrogen atoms bonded to a sp^3-C, but the -CH$_2$- hydrogens are more "*activated*" (lower C–H energy) due to the fact that the methylene is linked to two carbonyl (CO) moieties. CuO@GO has a light efficiency of *ca.* 10–13%, comparable to that of best microalgae in photobioreactors, [7] and affords two products, namely: CH$_3$COCH(CO_2H)COCH$_3$ (A) and CH$_3$COCH$_2$COCH$_2$$CO_2$H (B).

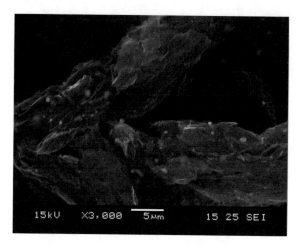

Fig. 10.5 Graphene oxide-GO loaded with CuO (bright spots) as photocatalyst for C–H carboxylation at room temperature under Vis-light irradiation. GO has itself some photocatalytic properties, much enhanced by CuO, and acts as CO$_2$ absorption medium and light adsorber

Table 10.2 Average C–H, C–C and C–O Bond energy as reference for the 2,3-dihydrofuran molecule

	$sp^3C\text{-}sp^3C$	$sp^3C\text{-}sp^2C$	$sp^2C\text{-}sp^2C$	$sp^3C\text{-}H$	$sp^2C\text{-}H$	$sp^3C\text{-}O$	$sp^2C\text{-}O$
Number of bonds	1	1	1	4	2	1	1
Average energy kJ/mol	346	386–390	610	411	422	358	418

Product (A) can be obtained also by thermal route using basic catalysts under high pressure of CO_2 and high temperature. Product (B), that implies the activation of more energetic -CH_3 moiety than -CH_2-, is unique of the photocatalytic route. So, solar-driven photocatalysis, under mild conditions (ambient P and T), can afford interesting reactions that the thermal route cannot afford.

2,3-Dihydrofuran (Fig. 10.6) is an interesting molecule that contains C–H bonds of different energy (Table 10.2): in practice the six hydrogens belong to four different energy levels as they are linked to either a sp^2- (C = C bond) or sp^3-C, and are either close or away from the O-atom [8].

The thermal carboxylation of 2,3-dihydrofuran-2,3-DHF (Fig. 10.6a) in the dark does not occur even at >100 °C in presence of Ru@ZnS-B (a structural variation of zinc sulphide modified with Ru nano-particles). When irradiated with Vis-light at ambient temperature in presence of the same catalyst, the organic species is carboxylated with a light efficiency close to 25% [8].

An alternative route of using photocatalysis is the reduction of CO_2. Nanosized CuI has been shown to be active in such process (b, in Fig. 10.6).

Noteworthy, TiO_2, a *n*-type semiconductor is not very efficient in this reaction, while CuI, a *p*-type semiconductor may afford different products, according to its size (Fig. 10.7).

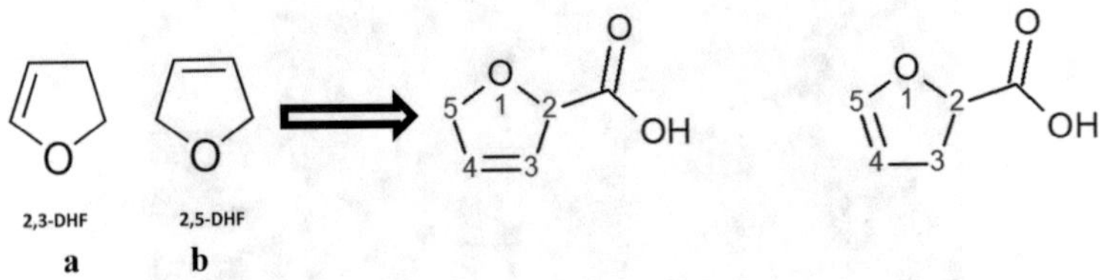

Fig. 10.6 Photocarboxylation of 2,3-DHF

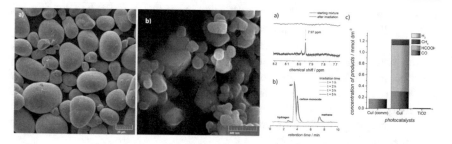

Fig. 10.7 TEM view of commercial (**a**) and nanosized CuI prepared in our laboratory. The former produces only a small amount of CO in the photochemical reduction of CO_2 in isopropanol under visible light irradiation, the latter produces HCO_2H with an interesting selectivity. The product distribution is shown on the right which also shows that ^{1}H-NMR and Gas-chromatography can be used to quantify the products as they are formed. Adapted from Ref. [8b] (CC BY 4.0)

As already mentioned, in order for a photocatalyst to work, its band gap (the difference in energy between the valence and conduction band (see Fig. 10.8)) must match the energy required for running the reaction. Figure 10.8 shows the band gap of some semiconductors (*n*- and *p*-type) and the potential reduction of CO_2 to a variety of products. TiO_2, very active under UV-light in the oxidation of organics, does not match the potential required for CO_2 reduction [9a].

"*Solar chemistry*" is of a paramount importance today as coprocessing of CO_2, and water using solar radiations is a way to the production of chemicals and fuels from non-fossil-C, avoiding the production, storage and transport of hydrogen. Particular interest has risen the bimetallic system such as SnO_2-coated Cu nanoparticles, which depending on the thickness of the tin oxide layer afford more selective CO (0.8 nm; >93% at -0.7 V_{VHE}) or formates (1.8 nm).

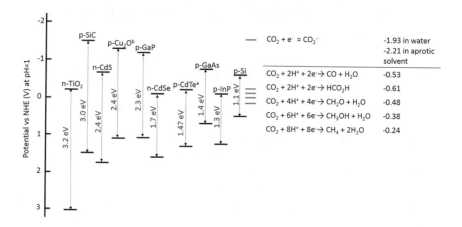

Fig. 10.8 Band gap (on the left) of some semiconductors and energy required in CO_2 reduction to selected products (on the right). Reprinted from Ref. [9a], Copyright (2016), with permission from Elsevier

Nanowires made of "SnO_2 on CuO" coupled with a GaInP/GaInAs/Ge photovoltaic cell show an interesting solar-to-CO conversion efficiency of 13.4% [9]. Using perennial energy for coprocessing CO_2 and water is of paramount importance for our future and for an extended C-recycling [10, 11]. Engineering new photomaterials such as mixed oxides, coupling a n-, and a p-type semiconductor is a way to avoid charge recombination and address the selectivity issue. Photocatalysts selectivity toward one of the reduction products of CO_2 is a key issue to solve. In fact, the energy necessary for product separation is a key factor in determining the overall energetics of a process.

Working in water, the ideal would be to produce a chemical which is not water-soluble, so that it can escape the condensed phase. Looking at Fig. 10.7, CH_4 and CO leave the water phase together with excess CO_2, while formic acid is water-soluble and remains in water. If only methane or only CO in the gas phase were produced even if coupled to formic acid, the product separation would be easier; viceversa, the separation of methane and CO eventually both present in the gas phase with CO_2, requires more energy for separation. Spending energy means today emit CO_2, as the use of PV is not yet at such economic level to drive the separation of industrial processes relevant to very cheap materials such as CO and CH_4. Therefore, CuI presents quite interesting photo properties but should be improved in order to cut one of the gaseous products.

10.5   PV-Driven Processes

The utilization of PV-electricity in CO_2 reduction can occur either in a direct or an indirect way. In the latter, PV-H2 is produced which is then used for CO_2 reduction, as already discussed in Chap. 9. In the former, CO_2 is directly used in electrochemical processes and converted into a variety of compounds. Moreover, the direct reduction can be performed in a variety of routes as described in the following paragraphs.

10.5.1   Electrochemical Reduction of CO_2

In an electrochemical device, electricity drives two processes occurring, respectively, at the cathode (reduction) and the anode (oxidation). Usually, the electrodes are immersed in a liquid which may even take part in the process. Working in water, it is worth to recall the low solubility of CO_2 and the fact that it can originate a variety of species, according to the pH. Hydrogen carbonate, HCO_3^- is most abundant and can be reduced instead of CO_2. The low solubility of CO_2 in water increases the energy necessary for the reduction, which often occurs at less negative potential than the thermodynamic ones. However, the pH of the solution must be kept under control.

The electrochemical reduction of CO_2 is performed in an electrochemical cell in which CO_2 is bubbled and reduced at the negative electrode (cathode). The electric energy is ideally generated using non-fossil sources (perennial SWHG, Nuclear). The electro-reduction can occur directly on the electrode surface (which can undergo corrosion or be consumed) or can be mediated by an "*electrocatalyst*" which accepts electrons from the cathode and transfers them to CO_2. In this way, the electrode corrosion and consumption are prevented. The electrocatalysts can be either attached on the surface of the electrode or dissolved in solution. By electrolysis, two different classes of products can be obtained:

i. Carboxylates: $CH_2 = CH_2 + 2CO_2 + 2e^- + 2H^+ = HO_2C-CH_2-CH_2-CO_2H$
ii. Reduced forms: $CO_2 \ggg CO, HCO_2H, CH_2O, CH_3OH, CH_4, Cn$

Coupled to the CO_2 reduction is the oxidation of an organic molecule or, better, of water to afford O_2. Figure 10.9 exemplifies a basic electrochemical cell. Many different modifications exist, some of them very sophisticated with electrodes adorned with complex metal systems, for improving yield and selectivity by tuning thermodynamic potential of CO_2 reduction reactions and applied voltage.

10.5.2 Photoelectrochemical Reduction of CO_2

In this technology, light is directly used in the electrochemical apparatus. It is shaded on photosensitive electrodes such as semiconductors (Fig. 10.10); electrons are produced through the irradiation and used in the reduction of CO_2. Obviously, an oxidation reaction is coupled to the reduction ($CO_2 \gg CO$) process. In Fig. 10.10, water is oxidized to O_2. In this case, there is no connection to the electric grid, neither PV systems are used for feeding electrons.

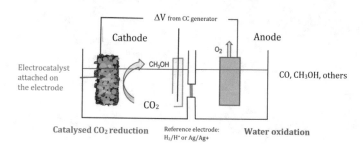

Fig. 10.9 Example of a basic electrochemical cell. The electrodes are connected to a source of continuous electric current and reactions occur at the electrode surface. In the specific case, electrocatalysts are attached on the cathode for selective CO_2 reduction (they can also be dispersed in solution). The electrocatalysts are metal complexes of living organisms or enzymes (see Chap. 11). Adapted from Ref. [8] (CC BY 4.0)

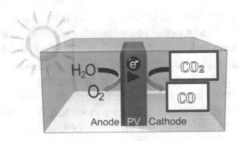

Fig. 10.10 Simplified photoelectrochemical cell. Electrodes are made of semiconducting materials. Upon irradiation, charge separation occurs, and electrons are circulated for the reduction of CO_2. No connection to the electric grid, nor to PV-cells is necessary

For a full exploitation of the Photoelectrochemical option, new materials are needed that assure a better light utilization efficiency coupled to improve catalytic properties and high selectivity. Such application has found quite innovative applications, as described in Chap. 11.

10.5.3 Electrochemical and Photoelectrochemical Processes Coupled with Biosystems

Such innovative application will be discussed in detail in Chap. 11. Here we just announce that the novelty is in the use of biosystems (enzymes or full microorganisms) as the "catalysts" in the reduction process. Both enzymes and

Table 10.3 Solubility of gases in water at 25 °C

Gas	Composition of gas in atmosphere (%)	Henry's constant (K_H) at 25 °C. (L* atm/mol)	Concentration in water (solubility)	
			(mmol/L)	(mg/L)
Nitrogen (N_2)	78.08	1639.34	0.48	13.34
Oxygen (O_2)	20.95	769.23	0.27	8.71
[a]Carbon Dioxide (CO_2)	4.10×10^{-2}	29.41	0.62	27
Neon (Ne)	1.82×10^{-3}	2222.22	8.18×10^{-3}	0.17
Helium (He)	5.24×10^{-4}	2702.70	1.94×10^{-6}	7.76×10^{-6}
Hydrogen (H2)	5.50×10^{-5}	1282.05	4.29×10^{-7}	8.65×10^{-7}

[a]Note that the value reported for CO_2 is relative to the "free CO_2" and does not include the amount of its hydrated forms such as H_2CO_3, HCO_3^-, and CO_3^{2-}

microorganisms can be attached to the cathode [12]. The reaction product is either HCO_2^- or $CH_3CO_2^-$ that can be further biotechnologically converted into longer chain hydrocarbons, reaching the use of products as fuels.

In some cases, hydrogen is produced photoelectrochemically which is used by microorganisms in solution. This helps to generate "in situ" a "*high concentration*" of H_2 that would not be otherwise generated by bubbling H_2 due to its low solubility in water (Table 10.3) [13]. Such hydrogen is used by microorganisms for the reduction of CO_2.

10.6 High-Temperature Processes Driven by Solar Power Concentrators

In Chap. 5, we have shown that it is possible to concentrate the solar energy and reach quite high temperatures, even above 1000 °C. Such temperature is used to drive energy-requiring reactions such as the CO_2 dissociation ($CO_2 >> CO + 1/2\ O_2$) and water splitting ($H_2O >> H_2 + 1/2O_2$). In order to improve the reaction rate and yield, the reactions are carried out in presence of a catalyst. The sequence of steps is represented in Eqs. 10.1–10.4.

$$MO_x > > MO_{x-1} + 1/2O_2 \tag{10.1}$$

$$H_2O + MO_{x-1} > > H_2 + MO_x \tag{10.2}$$

$$CO_2 + MO_{x-1} > > CO + MO_x \tag{10.3}$$

$$mH_2 + nCO > > CH_3OH,\ HCn \tag{10.4}$$

The oxide (MO_x) must be characterized by having a low lattice energy so that one oxygen atom can be easily dissociated leaving vacancies (vacant sites) in the lattice (MO_{x-1}). Such new oxide is a strong "oxophile" (is avid of oxygen) and when contacted with water (or CO_2) takes one oxygen and converts back into the

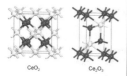

The red spheres are O, the yellowish are Ce. In order to identify the Ce-oxide units in the unit cell, one has to know that: one atom in the body of the cell belongs to that cell, one atom on face belongs ½ to the cell, one on the edge belongs for ¼ to the cell, one on the corner belongs for 1/8 to the cell. Therefore, in CeO_2 there are 8 red O within the unit cell. The Ce atoms come: 8/8=1 Ce from the corners plus 6/2=3 Ce from the edges, for an overall of 4 Ce: this makes that the unit cell contains 2 CeO_2 units. In the Ce_2O_3 unit cell, there are 2 Ce and 2 O atoms in the body of the cell plus 8/8=1 O form the corners. In total 2Ce and 3O: the unit cell contains one Ce_2O_3 unit.

CeO$_2$ Ce$_2$O$_3$

Fig. 10.11 The unit cell of CeO_2 (left) and Ce_2O_3 (right). Reprinted from Ref. [14] (CC BY 4.0) Copyright (2017) Indonesian Mining Journal

original oxide (MO_x) converting water into H_2 and CO_2 into CO. CeO_2 (cerium di oxide used for residual fuels, CO and NO conversion in catalytic converters in cars) (Fig. 10.11) finds application in such process. It can release one oxygen per two CeO_2 oxide units and converts into Ce_2O_3. The latter converts back to CeO_2 taking one O-atom from water or CO_2.

It is worth to emphasize that the loss of oxygen may not be stoichiometric, which means that the loss of oxygen may be much lower than 1 per two CeO_2 units. Oxygen can, thus, be lost in sub-stoichiometric amounts, by producing seldom vacancies in the lattice. In order to make easier the loss of oxygen, CeO_2 can be modified in several ways by adding other oxides which modify the original lattice structure and bond energy. Hydrogen and CO are used in the FT process to produce liquid hydrocarbons or can be converted into methanol, as we have already seen. As both CO_2 and $H_2O_{(vapor)}$ can be captured from the atmosphere, the product is named *Diesel from air* [15].

References

1. Ciamician G (1912) La fotochimica dell'avvenire, VIII international congress on applied chemistry, Washington and New York, 4–13 September 1912, 28:135–150
2. a) Armaroli N, Artero V, Centi G, Dibenedetto A, Hammarstroem M, Mul G, Pickett C, Rau S, Reek NH (2016) Solar driven chemistry. EUCheMS; b) Aresta M, Dibenedetto A, Angelini A (1996) The use of solar energy can enhance the conversion of carbon dioxide into energy-rich products: stepping towards artificial photosynthesis. Phil Trans Royal Soc A Math Phys Eng Sci 371:20120111
3. a) Andaloro A, Mazzucchelli ES, Lucchini A (2016) Photocatalytic self-cleaning coatings for building façade maintenance. Performance analysis through a case-study application. J Facade Design Eng 4:115–129; b) https://www.fondriest.com/environmental-measurements/parameters/weather/photosynthetically-active-radiation/
4. https://solarlight.com/product/xps-300-xenon-lamp-power-supply/
5. Hawecker J, Lehn JM, Ziessel R (1983) Efficient photochemical reduction of CO_2 to CO by visible light irradiation of systems containing $Re(bipy)(CO)_3$ X or $Ru(bipy)_3{}^{2+}$-Co^{2+} combinations as homogeneous catalysts. J Chem Soc Chem Commun 9:536–538
6. a) Hori Y, Wakebe H, Tsukamoto T, Koga O (1994) Electrocatalytic process of CO selectivity in electrochemical reduction of CO_2 at metal electrodes in aqueous media. Electroch Acta 39 (11–12):1833–1839; b) Hori Y (2008) Electrochemical CO_2 reduction on metal electrodes. In: Modern aspects of electrochemistry, 3rd edn. Vayenas, pp 42, 89–189
7. Dibenedetto A, Zhang J, Trochowski M, Angelini A, Macyk W, Aresta M (2017) Photocatalytic carboxylation of CH bonds promoted by popped graphene oxide (PGO) either bare or loaded with CuO. J CO_2 Util 20:97–104
8. a) Aresta M, Dibenedetto A, Baran T, Wojtyła S, Macyk W (2015) Solar energy utilization in the direct photocarboxylation of 2,3-dihydrofuran using CO_2. Faraday Discuss. 183:413; b) Baran T, Wojtyła S, Dibenedetto A, Aresta M, Macyk W (2016) Photocatalytic carbon dioxide reduction at p-type copper (I) iodide. ChemSusChem 9(20):2933–2938
9. a) Aresta M, Dibenedetto A, Quaranta E (2016) State of the art and perspectives in catalytic processes for CO_2 conversion into chemicals and fuels: the distinctive contribution of chemical catalysis and biotechnology. J Catalys, 343:2–45; b) Schreier M, Héroguel F, Steier L, Ahmad S, Luterbacher JS, Mayer MT, Luo J, Graetzel M (2017) Solar conversion of CO_2 to CO using Earth-abundant electrocatalysts prepared by atomic layer modification of CuO. Nature Energy 2(7):17087–17096

10. Zhang W, Hu Y, Ma L, Zhu G, Wang Y, Xue X, Chen R, Yang S, Jin Z (2017) Progress and perspective of electrocatalytic CO_2 reduction for renewable carbonaceous fuels and chemicals. Adv Sci 5(1):1700275–1700279

11. Aresta M, Nocito F, Dibenedetto A (2018) What catalysis can do for boosting carbon dioxide utilization. Adv Catal 62:49–110

12. Schlager S, Dibenedetto A, Aresta M, Apaydin DH, Dumitru LM, Neugebauer H, Sariciftci NS (2017) Biocatalytic and bio-electrocatalytic approaches for the reduction of CO_2 using enzymes. Energy Technol 5:1

13. https://www.molecularhydrogeninstitute.com/concentration-and-solubility-of-h2

14. Wahyudi T (2015) Reviewing the properties of rare earth element-bearing minerals, rare earth elements and cerium oxide compound. Indones Mining J 18(2):92–108

15. www.climeworks.com

Enhancing Nature

11

Abstract

This Chapter provides an overview of possible solutions to convert carbon dioxide into various bioproducts, such as biofuels, bioplastics and bio-sourced chemicals, mimicking Nature, or even enhancing natural systems. Biotechnological techniques or hybrid systems, made by integrating chemical(electro)-catalysis and biological systems, are discussed.

11.1 Introduction

The world leaders are committed to implementing global international agreements on climate change that may limit global warming to below 2 °C with respect to 1990. Nature may help to find solutions, as it is full of clues for how we find solutions to the climate change problems. Mimicking and enhancing Nature may help to convert large volumes of CO_2 even at a rate higher than that of natural system and with a better selectivity. Plants have not optimized the use of solar energy, because there is no need, as it is plentiful; they build-up a variety of compounds, either functional or structural or even energy-storage. Industrial processes optimize the use of energy toward a single product; by reading natural systems we can get inspiration for new ideas that may enhance natural processes.

In Nature one can find very intriguing systems dealing with CO_2 conversion either into chemicals or into renewable energy. For example, the *Saguaro cactus* (Fig. 11.1) can capture CO_2 from the atmosphere to make compounds called oxalates. Calcium oxalate (CaC_2O_4) is reported to be present in different families of plants, [1] and in such cases it represents more than 70% of their dry weight. Interestingly, the metabolism of oxalate by oxalotrophic bacteria allows the precipitation of calcium carbonate ($CaCO_3$) in acidic tropical soils, which are, otherwise, free of primary carbonates. One mole of calcium oxalate is catabolized into

© Springer Nature Switzerland AG 2021
M. Aresta and A. Dibenedetto, *The Carbon Dioxide Revolution*,
https://doi.org/10.1007/978-3-030-59061-1_11

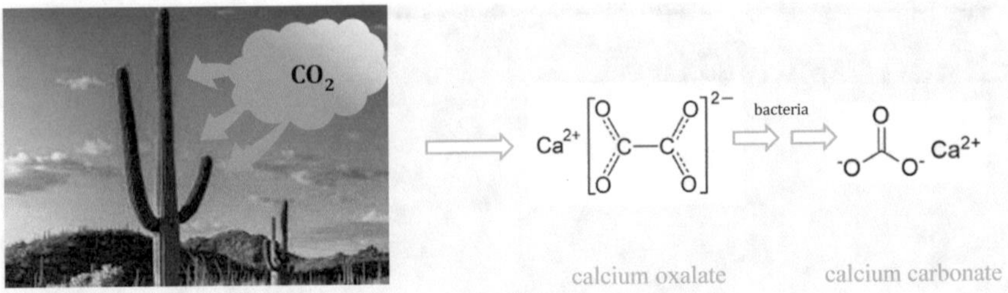

Fig. 11.1 *Saguaro Cactus* and sequestration of CO_2 from atmosphere into soil

one mole of carbonate with the release of one mole of respired carbon dioxide. Fungi also assist in the process, by either breaking down oxalic-rich matter and depositing calcium oxalate for catabolism by bacteria or by fungal oxalotrophy.

This process is central to the oxalate–carbonate pathway (Fig. 11.1), which couples the biogeochemical cycles of calcium and carbon, and is gaining increasing interest as a potential long-term sink for atmospheric CO_2 [2, 3]. The formation of oxalates from CO_2 has been targeted by chemists for a long time; it is not a simple process because it implies a one-electron transfer to CO_2, a high energy process, and coupling of short-living CO_2^- radicals. Plants do it quite simply!

As it has been reported in the previous Chapters, scientists are seeking a number of ways for carbon sequestration and utilization, developing clean technologies, and energy management innovations for reducing the immission of CO_2 into the atmosphere. Some companies target the sequestration of CO_2 into useful polymers avoiding toxic substances and mimicking Nature. Others produce cement mimicking hard corals (Fig. 11.2) which use CO_2 and calcium to biomineralize their skeleton in seawater [4]. Also, energy management systems are developed mimicking Nature, inspired by how groups of organisms use simple rules to self-organize, producing collective, "intelligent" behaviors.

Fig. 11.2 Hard corals
Staghorn coral (*Acropora
cervicornus*) in Cordelia
Banks

Fig. 11.3 Peak energy demand: the orange line outlines the highest level of demand in a given day/month; the green area highlights as the wireless network is able to optimize the energy demand with peak shaving

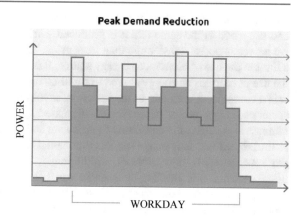

Peak Demand Reduction

POWER

WORKDAY

Based on this, companies aim at developing systems that are able to wirelessly connect building's power-consuming appliances to each other, enabling them to detect each other's power cycles, and communally determine the best times to turn each appliance on and off. The result is a lower, smoothed-out energy demand (Fig. 11.3), which reduces both costs for building owners and strain on central electrical grids during peak demand [5].

Nature is very inspiring not only for the number of processes amenable to produce chemicals, energy products, and new materials, but also for its fundamental paradigm to *not produce waste.*

Terrestrial plants, algae, and bacteria are photosynthetic organisms, so they can capture and convert sunlight into valuable products, storing energy into chemical bonds. Through photosynthesis, these organisms may live and grow, using carbon dioxide and water and release oxygen, useful for life on Earth. Science and Technology can try to improve natural processes in many different ways (including genetic modification of microorganisms that is not accepted in several countries) or can mimic natural processes and transfer them into industrial plants that are more intensive than Nature, and with a better selectivity toward a target product. Based on the above considerations, using *Carbon Dioxide, Water, and Sunlight as energy input, a new economy can be built* [6]. In following paragraphs, a number of cases will be discussed trying to highlight opportunities that are either inspired by Nature or even try to improve natural performances.

11.2 Direct and Indirect Use of Biomass as Source of Energy and the Enhanced CO₂ Fixation

Biomass can be used to produce energy as alternative to fossil fuels. A positive fact is that plants are able to capture almost the same amount of CO_2 through photosynthesis while growing as what is released when biomass is burned; this makes biomass a quasi-carbon-neutral energy source. So, in principle, we could burn wood for energy production with zero-CO_2 emission; the must is that we keep a balance and use wood at the same rate it is produced. Unfortunately, this is not the case as we harvest wood at a faster rate than trees grow. This is the big problem, which sets a limit to the use of biomass as energy source. It is useful to recall that the shift from the use of biomass as energy source to fossil-C, more dense in energy, in late 1700s was a compulsory step due to the industrial revolution. Biomass cannot satisfy our society's energy needs. Moreover, as discussed in Chap. 5, the direct burning of biomass also causes the emission of particulate and noxious pollutants. So, we have to find routes that may cause less environmental burden, such as the conversion of the raw materials into biofuels. Additionally, taking into account the carbon up-taken from soil, the use of biomass can become a net transfer of CO_2 from surface soil to the atmosphere, unless clever strategies are implemented. Using biomass, thus, is not "per se" a way to a zero-emission of CO_2, and it will reach such target if we use soil with the right strategy and apply the concept of biorefinery to the use of biomass. Ethanol and biodiesel can be valid alternatives to fuels derived from fossil-C supposed that the conflict between food-energy and the use of soil is solved. Biogas, which is formed by fermentation of fresh organic waste or fresh non-eatable grown biomass, is composed mainly of methane (60%) and CO_2 (40%), which is an interesting option of using waste or non-eatable biomass. However, the winning strategy is the use of waste biomass, biomass grown on marginal soils (such as polluted or arid soils, which are not suited for growing edible biomass), and the use of non-edible biomass.

11.2.1   Production of Fuels and Chemicals from Aquatic Biomass

Algal technology is attracting a lot of attention for the ability of algae to fix inorganic carbon and convert it into several value-added products, including biofuels precursors. Both macro- and microalgae are under analysis for exploitation. Various strains have a different potential, show a different adaptation to being grown out of their natural environment, and can be useful to produce high value products, materials, and fuels for market [7, 8].

Macro- and microalgae are rich in protein, carotenoids, antioxidants, fatty acids, enzymes, polymers, peptides, toxins, and sterols and other compounds, which can be used for several purposes (Fig. 11.4).

Microalgae have been identified as a third generation feedstock, and a more efficient source of biodiesel (estimated production: 50–70 t ha^{-1} y^{-1} in open ponds and 150 t ha^{-1} y^{-1} in photobioreactors) compared to terrestrial energetic oily crops (~ 3 t ha^{-1} y^{-1} for soybeans, ~ 9–15 t ha^{-1} y^{-1} for corn, and ~ 13 t ha^{-1} y^{-1} for switch grass) [9]. Nevertheless, due to their complex composition (Fig. 11.4) the use of microalgae for making only biodiesel is not economic [10]. However, it is necessary to apply the concept of biorefinery [7] for a most viable economic use of algal biomass [11, 12].

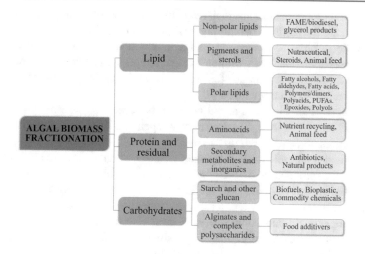

Fig. 11.4 Chemicals derived from algae [7b]

Microalgae can be grown in different cultivation systems: either open ponds (Fig. 11.5a) which are more economically acceptable or photobioreactors (PBRs) (Fig. 11.5b) which are more expensive.

The production of microalgae in open ponds strongly depends on solar irradiation, temperature, and contaminants, which may affect the productivity and composition. Such parameters together with the need of nutrients, the availability of land and water, drive the economics of the process. The fact that microalgae do not require soft water for growing, but can be grown in ocean water, or else in sanitized process- and waste-municipal-waters, is a big point in their favor with respect to land biomass and contributes somehow to lower the production costs.

Fig. 11.5 Cultivation of microalgae in: **a** open pond; **b** photobioreactors-PBRs [13]

Other important operations that may influence the economics of microalgae production are harvesting [14], drying, and processing of biomass. A huge effort is being made in order to reduce the energetic cost and CAPEX of equipment necessary for separation and dehydration of algal biomass, in addition to implementing efficient fractionation techniques for efficiently applying the concept of biorefinery and valorize all algal components [7].

Microalgae (Fig. 11.6) can afford not only biodiesel, but also biogas as well as bioethanol, biohydrogen and hydrocarbons depending on their composition. So, a lipid rich biomass (>40% dw, up to 70–75% dw) will be useful for the production of bio-oil and biodiesel, while a biomass rich in carbohydrates [15] will be better suited for the production of bioethanol. The anaerobic fermentation of sugars, proteins, lipids, and organics will produce biogas, a process that for some strains rich in proteins may require a careful management of ammonia elimination for an optimal biogas yield and composition. It must also be pointed out that achieving high concentration of lipids (70^+%) is possible in the lab, much less in extended cultures in PBRs, and practically impossible in open ponds because of the negative influence of physical parameters and parasites or contaminants. In order to increase the competitiveness of microalgal biomass [16, 17], it is necessary to utilize all their components implementing the *algal fractionation* (Fig. 11.4) and applying the concept of biorefinery in the conversion of biomass into sustainable fuels and added value chemical products with quasi-zero waste production.

Moreover, it is worth to say that lipids derived from algal biomass usually are not made by a single predominant fatty acid-FA (as occurs in drupes of some land plants, such as palms that contain 90+% of saturated palmitic acid), but are frequently constituted of several different FAs (saturated, monounsaturated, di-unsaturated, polyunsaturated, including omega-3) esterified with glycerol. Such lipids are not all suited for conversion into biodiesel (esters with more than one

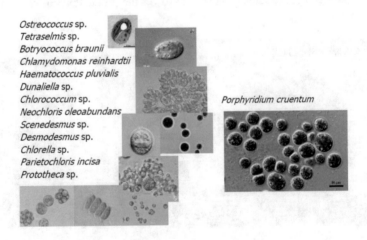

Ostreococcus sp.
Tetraselmis sp.
Botryococcus braunii
Chlamydomonas reinhardtii
Haematococcus pluvialis
Dunaliella sp.
Chlorococcum sp.
Neochloris oleoabundans
Scenedesmus sp.
Desmodesmus sp.
Chlorella sp.
Parietochloris incisa
Prototheca sp.

Porphyridium cruentum

Fig. 11.6 Selected strains of microalgae cultivated on a large scale

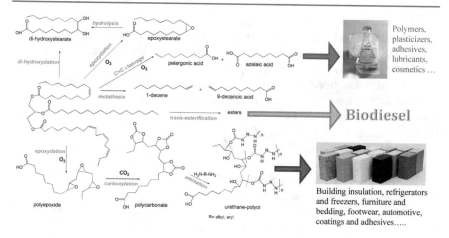

Fig. 11.7 Special uses of lipids according to their nature

unsaturation can cause problems to engines during combustion) and need treatment (hydrogenation) or should be separated in their components, as shown in Fig. 11.7, each used for several different applications [18].

The cost of microalgae production (growth, harvesting, and drying) has been evaluated at 1.1 €/kg (non-optimized, experimental value) [12] with 0.68 €/kg as best optimized, theoretical cost. Even a cost of 0.40 €/kg has been proposed in the most optimistic of cases. If algae are used to produce only biodiesel, assuming a good lipid concentration of 35%, their value would be 0.17 €/kg, which is much lower than current production cost. Conversely, if fractionation of microalgae is carried out and proteins, carbohydrates, and the various lipids used each for their best use (Fig. 11.7), then a remunerative value of 1.65 €/kg can be reached.

Growing and exploitation of algal biomass demand great expertise and continuously updated competence. A great effort is being made for growing microalgae rich (>40%) in selected lipids.

11.2.2 Production of Fuels and Chemicals from Terrestrial Biomass

Terrestrial biomass depending on its characteristics can be categorized as lignocellulosic biomass, fresh vegetables, and oily biomass, which can be used for different purposes. Lignocellulosic biomass is, in general, used to produce biofuels, mainly bioethanol. It is composed of carbohydrate polymers (cellulose and hemicellulose) and lignin (an aromatic polymer) (Fig. 11.8).

Such polymeric materials are first depolymerized into Cn-polyols, used as commodities or platform molecules (Fig. 11.9) [19, 20] to produce fuels, chemicals, goods, by using different technologies (thermal processes, chemical catalysis, enzymatic catalysis, and hybrid systems).

The conversion processes can be categorized as therm(ochemic)al, biochemical, and chemical processes. Very often, lignocellulosic material is heated (temperature above 180 °C) in high-pressure steam in order to separate it into its components (cellulose, hemicellulose, and lignin).

Such process is called Steam-Explosion (SE). The separated fractions can be treated by using various technologies to obtain fuels, chemicals, and advanced materials. Figure 11.10 summarizes the application of various technologies to the conversion of lignocellulosic biomass.

Pyrolysis consists of heating lignin or cellulose in anaerobic conditions at high temperature (975–1050 °C) for its conversion into a gas phase (10–20%), a liquid (bio-oil) (60–75%), and char (15–25%). Pyrolysis is a way for concentrating energy, converting low-density biomass into higher density bio-oil [21], which would more conveniently be transported to large treatment plants.

Gasification is a thermal process that converts biomass into a gaseous fuel mixture (H_2, CO, CO_2, CH_4, and some C_{2+} hydrocarbons) in the controlled presence of an oxidant (air, oxygen) and water vapor. The gasification of biomass can be performed in fixed, moving, or fluidised bed reactors at high temperatures (670–850 °C) [22].

Biochemical processes that use microorganisms are gaining popularity due to their financial and environmental benefits [23]. The production of bioethanol, obtained via microbial fermentation of depolymerized cellulose, has reached in 2018 a total volume of about 107.7 Mm^3 (Fig. 11.11) [24].

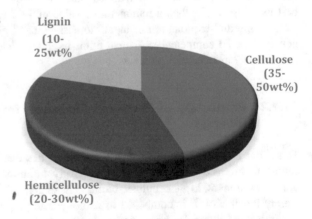

Fig. 11.8 Major components of lignocellulosic biomass

Lignin (10-25wt%

Cellulose (35-50wt%)

Hemicellulose (20-30wt%)

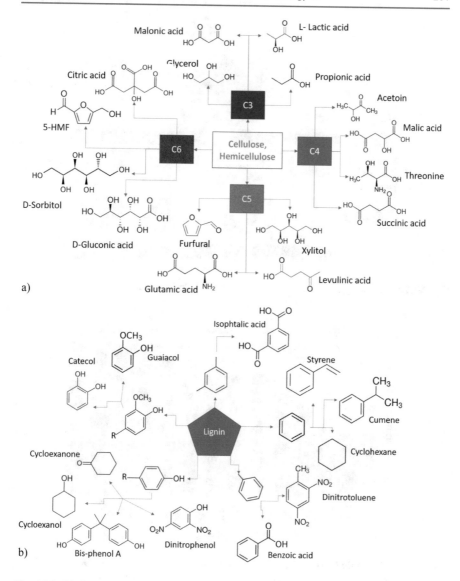

Fig. 11.9 Platform molecules derived from cellulose–hemicellulose (**a**) and lignin (**b**)

Alternatively, as already mentioned, cellulose and hemicellulose (Fig. 11.9) can be transformed into platform molecules (Fig. 11.12) such as furfural (2-furaldehyde), 5-hydroxymethylfurfural (5-HMF), and levulinic acid, which are intermediates for the production of a variety of chemicals or fuels.

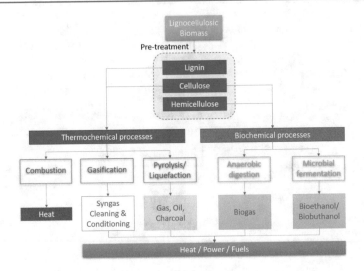

Fig. 11.10 Technologies for the conversion of lignocellulosic biomass

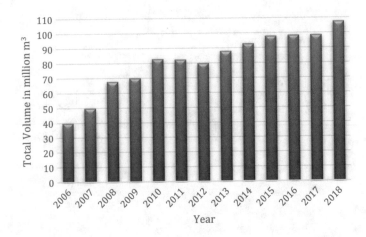

Fig. 11.11 The world bioethanol production: it has more than doubled from 2006 until 2018

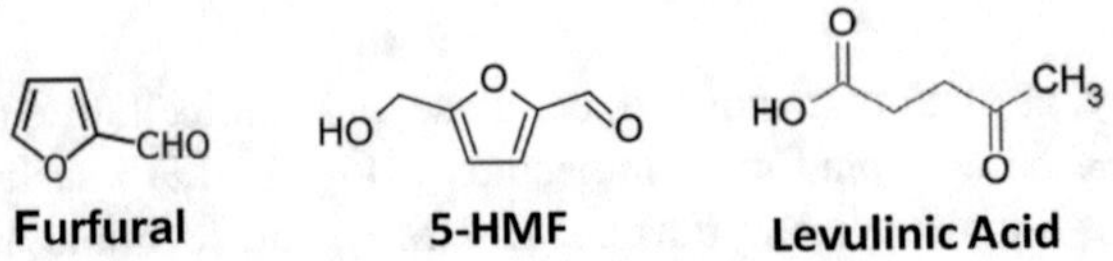

Fig. 11.12 Platform molecules derived from cellulose and hemicellulose

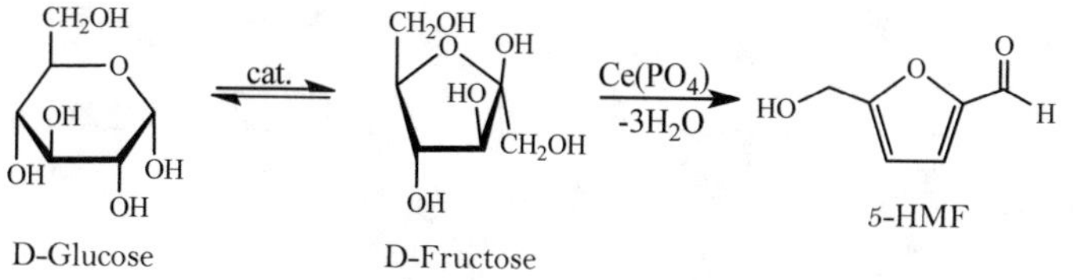

Scheme 11.1 Synthesis of 5-HMF by isomerization–dehydration of glucose

The conversion of glucose into 5-HMF is more effective if it is first isomerized into fructose (basic catalysis) which is then dehydrated using different homogenous or heterogeneous acid catalysts [25–32] under different reaction conditions (Scheme 11.1). In order to maximize conversion and selectivity, dimethyl carbonate (DMC) and its congeners have been used as extraction solvents of 5-HMF from an aqueous medium, using cerium (IV) phosphates as dehydration catalysts [33, 34]. DMC can almost quantitatively be recovered and reused.

A variety of biomass can be used for the synthesis of 5-HMF. For example, AVA Biotech HTC uses syrups extracted from energy crops to produce 5-HMF at various levels of purity (up to 99%) [35]. Avoiding the two-step process (isomerization–dehydration) would shorten the way to 5-HMF, and therefore there is an important incentive to develop catalysts that may directly drive the one-pot transformation of glucose into 5-HMF with high yields.

Due to the presence of the alcoholic and aldehydic moieties in its structure, (Fig. 11.13) 5-HMF can undergo oxidation processes affording a variety of derivatives (Scheme 11.2). Each of the oxidation products shown in Scheme 11.2 has a specific industrial application, as fine chemical, intermediate or monomer for polymers. Different reaction conditions have been applied using a variety of catalysts (homogeneous and heterogeneous), trying to minimize the cost of catalysts, using cheap and Earth-crust abundant metals, recovering and reusing them, avoiding expensive oxidants and organic solvents in order to optimize the economic and energetic cost of their production. In particular, mixed metal oxides (cheap and easy to handle), oxygen or air as oxidant and water as solvent have been used,

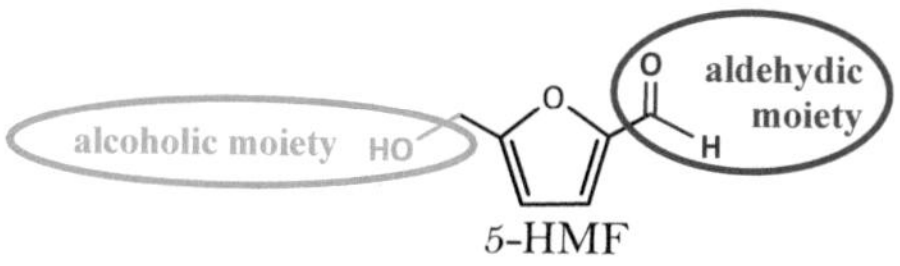

Fig. 11.13 Reactive functionalities of 5-HMF

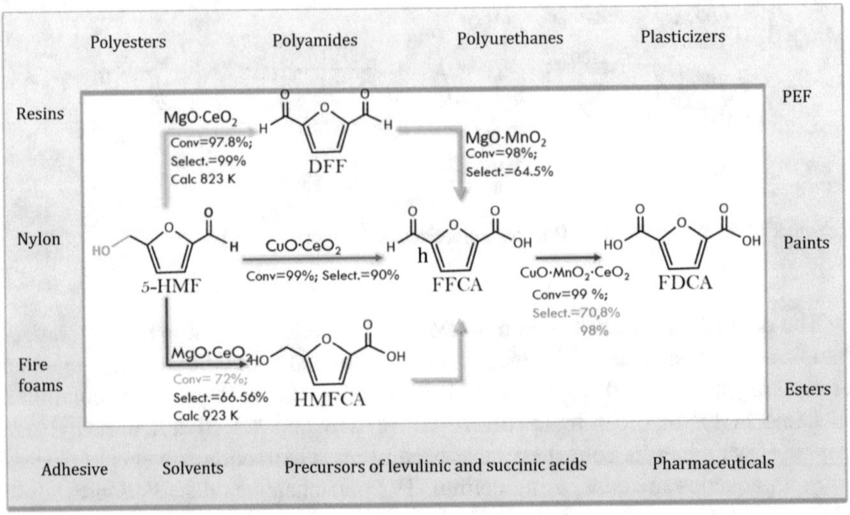

Scheme 11.2 Monomeric compounds obtained from HMF and their applications

Scheme 11.3 Oxidation
cleavage of HMF

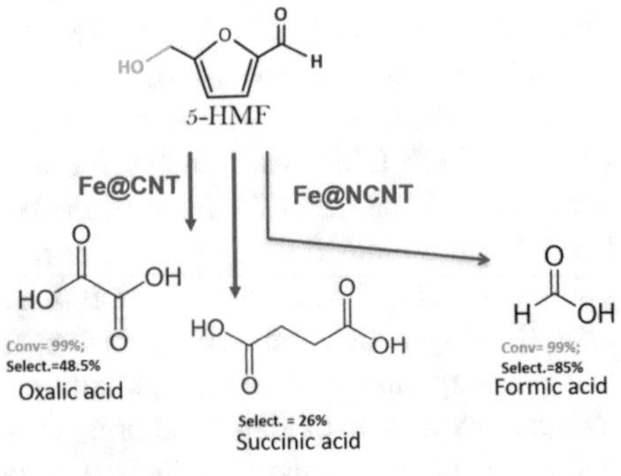

targeting high conversion yield and high selectivity toward products such as FFCA
[37], DFF [38], FDCA [39]. As depicted in Scheme 11.2, each derivative is produced with a yield and selectivity as close as possible to 100%.

Polyethylene furandicarboxylate (PEF)

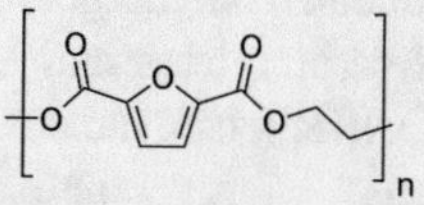

The value for 2,5-Furandicarboxylic Acid (FDCA) market was USD 391.3 million in 2018. The projected production potential by 2020 is 496.0 Mt, and a value of 498 MUS$, out of which 65% goes to PEF and the rest mainly to polyamides, polycarbonates, and plasticizers. The market is expected to reach 786.3 MUS$ by 2026 [36].

FDCA attracts much attention as it can be used to make polyethene furandicarboxylate (PEF), a potential effective substitute of petroleum-derived polyethene terephthalate (PET). Moreover, HMF can afford oxalic (OA) and succinic acid (SA) via aerobic oxidative cleavage [40] with high conversion and good selectivity (Scheme 11.3).

Another interesting derivative obtained by conversion of 5-HMF is levulinic acid (LA, 4-oxopentanoic acid) (Scheme 11.4) which then can be converted to γ-valerolactone (GVL) by catalytic hydrogenation at relatively low temperatures (100–270 °C) and high pressures (5.0–15.0 MPa), using either homogeneous or heterogeneous catalysts [41, 42].

Of great interest are also integrated systems that couple chemical and biological processes. To cite an example, Anbarasan et al. [43] have proposed a system where a fermentation process is combined with catalysis. The mixture of acetone–butanol–ethanol (ABE) obtained from glucose after the fermentative process is converted into C5–C15 fuels by using Pd/C-K$_3$PO$_4$ catalyst.

The use of enzymes allows the conversion of hemicellulose into monomers at mild conditions (pH 5, temperature 45–50 °C, atmospheric pressure) reducing the downstream efforts due to the use of acid (or base) chemicals required with chemical hydrolysis. Moreover, by using enzymes the selectivity is higher without the formation of byproducts and less energy consumption [44].

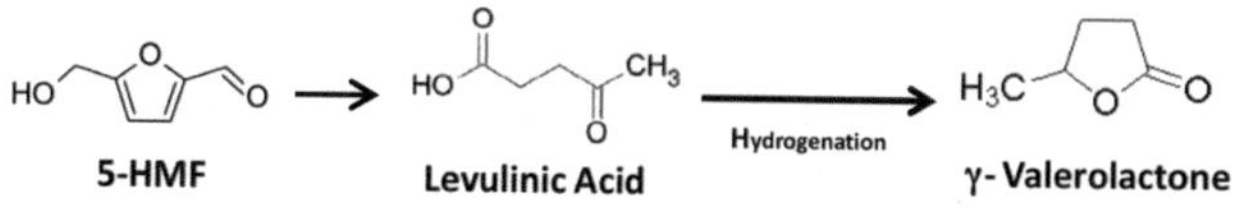

Scheme 11.4 Conversion of 5HMF to γ-valerolactone (GVL)

11.3   Carbon Dioxide and *Biotechnologies*

CO_2 can be used as C-1 building block for chemicals and materials or as source of carbon for the synthesis of products. The energetics of the chemical conversion of CO_2 says that the synthesis of carboxylates and lactones (RCOOR'), carbamates (RR'NCOOR''), ureas (RR'NCONRR'), isocyanates (RNCO), and carbonates [ROC(O)OR'] require moderate external energy input, if not zero [45, 46], while formates, methanol or methane and hydrocarbons require energy and hydrogen. Plants or microorganisms in Nature convert carbon dioxide and water into a large variety of energy-rich products using either solar energy or even chemical energy under ambient conditions.

Polyhydroxyalkanoates (PHAs)

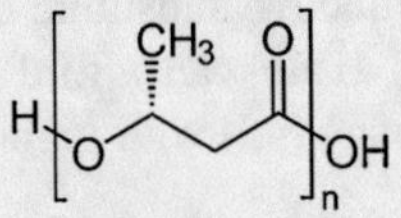

are biodegradable polymers for the production of bioplastics or biocomposites. They are either thermoplastic or elastomeric materials, with melting points ranging from 40 to 180 °C. The mechanical- and bio-compatibility of PHA can also be changed by blending and modifying the surface or combining PHAs with other polymers, enzymes, and inorganic materials, making it possible in a wide range of applications. The global polyhydroxyalkanoate market size is projected to reach 98 MUS$ by 2024, growing at a CAGR of 11.2% [49].

Cyanobacteria represent a good example of microbial platforms as they can use organic substrates and CO_2 to afford several useful compounds [47, 48]. They can easily be genetically manipulated and used for the production of quite different classes of products, such as fuels or polyhydroxyalkanoates (PHA), both made from carbon dioxide or in a mixed regime where an external organic substrate is also provided. Noteworthy, genetic manipulation is not accepted in several countries. Alternative physical stress technologies can be used to address the production of bioproducts.

2,3-Butanediol (2,3-BD)

is an important chemical used for manufacturing rubber and polyurethane commodities. It is also an intermediate chemical for the production of buta-diene, butene, and methyl ethyl ketone (MEK). In terms of value, the global 2,3-butanediol market is anticipated to expand at a CAGR of ~3%, to reach a value of ~220 MUS$ by 2027 [50].

Native and heterologous pathways in cyanobacteria produce several chemicals such as 1,3-propanediol (1,3-PD), 2,3-butanediol (2,3-BD), 2-methyl-1butanol (2-MB), 2-phosphoglyceraldehyde (2-PGA), 3-hydroxyproprionic acid (3-HP), 3-phosphoglyceraldehyde (3PGA), and also ethanol, isopropanol, buthanol, lactic acid, and other compounds, each with a different production yield (Table 11.1) depending on the used strain.

Recently, *Synechococcus elongatus* PCC 7942 has been engineered to convert CO_2 to fatty acid ethyl esters (FAEEs) that may find a use as biodiesel [51]. Several companies are using microorganisms for the direct conversion of CO_2 to respond to market requirements. The Dutch company Photanol is producing alternative products in the flavor and fragrance market [61], while Phytonix [62] produces n-butanol which can be converted into fuels and chemicals, including jet fuels, bioplastics, and synthetic rubber. NOVAMONT since 2016 is producing 1,4-butanediol (1,4-BDO) on an industrial scale directly from polyols and using bacteria at competitive price with fossil-C derived 1,4-BDO, [63] used as a solvent

Table 11.1 Chemicals produced by cyanobacteria

Organism	Product	Titer (g/L)	Productivity (mg/L·d)	References
Synechocystis sp PCC6803	Ethanol	5.5	212	[52]
	Lactic acid	0.8	60	[53]
	3-Hydroxypropionic acid	0.8	139	[54]
Synechococcous elongatus PCC7942	Isopropanol	0.03	3	[55]
	1,3-Propanediol	0.3	21	[56]
	1-Butanol	0.4	51	[57]
	2-Methyl-1-butanol	0.2	16	[58]
	2,3-Butanediol	12.6	1100	[59]
	Isobutyraldehyde	1.1	150	[60]
	Isobutanol	0.5	75	[60]

and for the production of plastics, elastic fibers and polyurethanes. The global 1,4-BDO market size was valued at USD 6.19 billion in 2015 and is expected to grow at an estimated CAGR of 7.7% from 2016 to 2025 [64].

Another example of biotechnological application comes from the UK company Deep Branch Biotechnology that, together with Drax Power Station, the biggest renewable electricity generator in the UK, has developed the production of proteins (titre 70%) from CO_2 by using a pool of microorganisms [65].

All cases presented above are good examples of application of the principles of the circular economy.

11.4   Bioelectrochemical Systems

Bioelectrochemical systems (BES) have recently been proposed as a new and sustainable technology for the generation of energy and useful products from waste [66]; in a BES, bacteria interact with solid-state electrodes (covered with semi-conductor materials so that electricity is generated by shining light on the electrode surface) and exchange electrons with them, either directly or via redox mediators. Microbial fuel cells (MFCs) and microbial electrolysis cells (MECs) are examples of BES (Fig. 11.14).

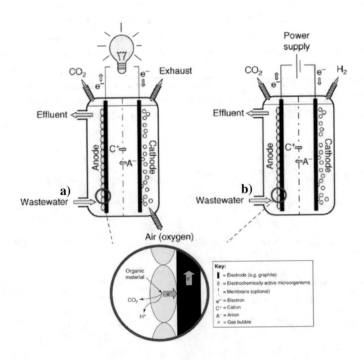

Fig. 11.14   a Microbial fuel cells (MFC); **b** microbial electrolysis cell (MEC). Reprinted from Ref. [67], Copyright (2008), with permission from Elsevier

A microbial electrolysis cell (MEC) is a technology related to microbial fuel cells (MFC). While MFCs produce an electric current from the microbial decomposition of organic compounds, MECs partially reverse the process to generate hydrogen or methane from organic materials by applying an electric current possibly generated from perennial sources.

In a basic bioelectrochemical system, electrons are generated at the anode from water or organic waste and sulfides [68], while microorganisms perform CO_2 reduction to organics at the cathode. The advantage of using BES is their productivity which is higher (5–7% efficiency) than photo-biological systems (3–4% solar conversion efficiency) [69].

By coupling BES with a photovoltaic device, the capacity of converting CO_2 using solar light may be considerably increased.

An interesting application of BES is the production of methane through electrochemical reduction of carbon dioxide at ambient conditions (Eq. 11.1).

$$CO_2 + 8H^+ + 8e^- \rightarrow CH_4 + 2H_2O \tag{11.1}$$

This latter process can be carried out using chemo-catalysis, using special-design reactors and catalysts with high activity under pressure (3 MPa) and at high temperature, despite it being a hexoergonic reaction. BESs target high selectivity and low overpotentials to be made applicable on a large scale. By using BES, the action of microorganisms is fastened by introducing electrons in the system that are used for reduction processes [70] of carbon dioxide that is converted into a variety of products. It can be reduced to methane, acetate, and other low-carbon products such as isobutanol and 3-methyl-1-butanol, depending on the used microorganism and the operational conditions [71].

Methane production has been reported by Villano et al. [72] using a hydrogenophilic methanogenic culture (Fig. 11.15) that reduces carbon dioxide to methane, at high rates (up to 0.055 ± 0.002 mmol d^{-1} mg VSS^{-1}) (VSS = Volatile Suspended Solids) and electron capture efficiencies (over 80%). The oxidation of organic waste is required in order to make both electrons and CO_2 available. Power can be supplied by a PV-Cell.

The reduction of carbon dioxide to methane by BES catches the attention of several groups and several papers have recently been published [73–77].

Moreover, in a recent study [78] a techno-economic assessment of biomethane production has been presented using a CO_2-containing effluent generated applying the ABAD Bioenergy® [79] technology, which is applied at demonstration scale (TRL8–9) in a variety of Wastewater Treatment Plants (WWTPs). CO_2 is reduced at the cathode of a BES, while the anode reaction is valorized by producing chloro-derivatives (Fig. 11.16) used as disinfectant of the treated wastewater. The application of such methodology increases the biomethane production by 17.4% and produces enough chlorine compounds for water sanitization.

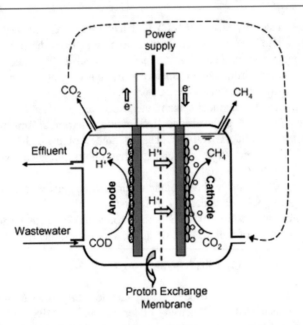

Fig. 11.15 Schematic drawing of a bioelectrochemical system for wastewater treatment and simultaneous CH_4 production based on bioelectrochemical CO_2 reduction. Reprinted from Ref. [72], Copyright 2010 with permission from Elsevier

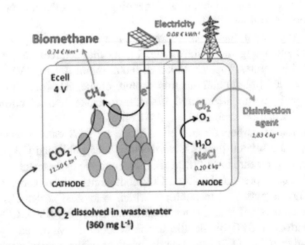

Fig. 11.16 Bioelectrochemical treatment of CO_2 dissolved in wastewater. Reprinted from Ref. [78], Copyright (2017), with permission from Elsevier

Bioelectrochemical systems have been used also for the microbial electrosynthesis (MES) to generate acetate from the biocathodic reduction of carbon dioxide with a continuous acetate production rate of 0.98 mmol C L_{NCC}^{-1} d^{-1} (NCC = net cathode compartment) under controlled pH-conditions (pH = 5.8) [80].

11.5 Integration of Catalysis and Biotechnology

An integrated approach that combines the strong points of electrochemical processes and biosystems in CO_2 reduction could be a way to take advantage of the high conversion rate and selectivity of metal enzymes intensified by using chemical processes. A "two-step" approach for CO_2 reduction to methane has been proposed. CO_2 is first reduced very selectively to formate electrochemically, then formate is separated by electrodialysis and used as feed by *Methanococcus maripaludis*, a methanogen microorganism [71]. Such kind of system combines chemo(electro)-catalysis with enzymatic catalysis, making use of solar energy for the selective conversion of CO_2 into useful, energy-rich products.

An integrated Photobioelectrochemical System (IPBES) has been used for the continuous conversion of CO_2 into formate [81] (Fig. 11.17).

A novel biocathode was setup, where enzymes and cofactors were inserted in a polydopamine (PDA) matrix and deposited on the surface of the cathode as thin films. The PDA matrix has a double function as it allows the immobilization of enzymes and its cofactors, and provides them with a suitable physiochemical environment so the enzymes remain stable for more than two weeks. The biocathode is then integrated with a visible light-driven anode photocatalyst, $BiVO_4$, that generates "electron-hole" pairs through irradiation and is coupled with the

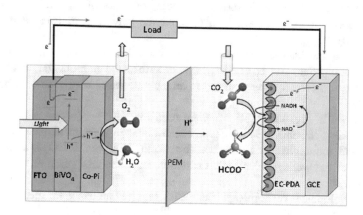

Fig. 11.17 Integrated photobioelectrochemical system (IPBES) for continuous conversion of CO_2 into formate. Reprinted from Ref. [81], Copyright (2017), with permission from Elsevier

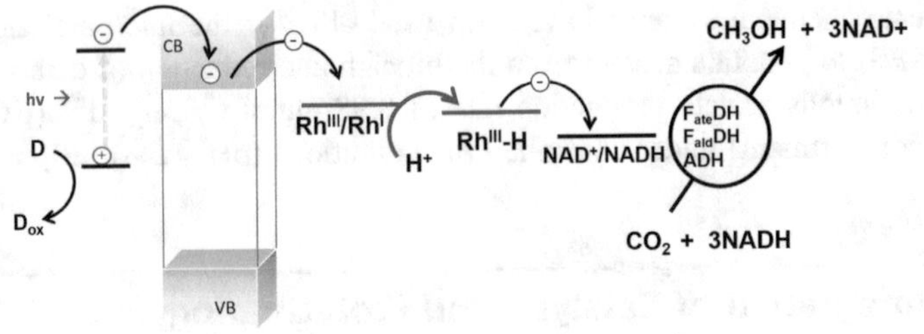

Fig. 11.18 Photoreduction of NAD + to NADH and integration in CO_2 reduction. Reprinted from Ref. [82] (CC BY 4.0)

oxygen evolution cocatalyst Co-Pi for effective use of water as source of $(H^+ + e^-)$ used in CO_2 reduction.

Of interest is also the hybrid bioinspired system used for the enzymatic reduction of CO_2 to CH_3OH in water [82] under ambient conditions, mediated by three NADH-dependent dehydrogenase enzymes co-encapsulated into a Ca-alginate matrix. Such *"dream reaction"* combines the enzymatic reduction of CO_2 to CH_3OH with the recycling of NAD^+ promoted by in situ photocatalytic reduction of NAD^+ to NADH, using semiconductors such as Cu_2O, $InVO_4$, TiO_2 modified either with the organic compound rutin or with the inorganic complex $[CrF_5(H_2O)]^{2-}$ [83] and the doped sulphide Fe/ZnS (Fig. 11.18).

The regeneration of the cofactor NADH from NAD^+, essential for an exploitation of the reaction, as three mol of NADH are used in the $6e^-$ reduction of one mol of CO_2, has been achieved by using visible light-active heterogeneous TiO_2-based photocatalysts.

The efficiency of the regeneration process is enhanced by using a Rh^{III}-complex for facilitating the electron and hydride transfer from the H-donor (water or a water-glycerol solution) to NAD^+. In this way, one mol of NADH was used for producing *ca.* 100 mol of CH_3OH, opening the way to a practical application (Fig. 11.20) [82].

Following the idea of catalytic electrodes for heterogeneous electrochemical CO_2 reduction, Schlager et al. [84] have described the immobilization of dehydrogenases and cofactor encapsulated in an alginate-based matrix on a carbon felt electrode.

As shown in Fig. 11.19, electrons are used for NAD^+ reduction which injects "$H^+ - e^-$" into the enzymes, which are encapsulated in alginate–silicate hybrid gel (green) and immobilized on a carbon felt working electrode. CO_2 is reduced to methanol at the working electrode. Oxidation reactions take place at the counter electrode.

The development of highly efficient BES requires an excellent electrical connection between the electrochemical system, the cathode, and the microorganism or enzymes, which may present some bottlenecks. For example, copper could be considered an excellent element for making a high conductive cathode, but its

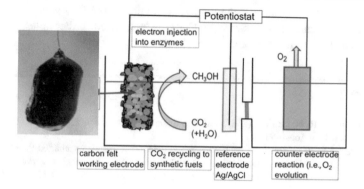

Fig. 11.19 Electrochemical cell for electroenzymatic CO_2 reduction Adapted from Ref. [84] (CC BY 4.0)

antimicrobial properties represent a limit for bioelectrochemistry. Bioelectrochemical reduction of CO_2 to acetate has been achieved using microbial electrosynthesis (MES) reactors. Pant et al. [85] have studied the production of acetate from CO_2 combining MES system with an anion exchange resin for sorption of acetate from the broth. Acetate desorption from the resin allows to have a final concentration of 5.0 g L^{-1} of acetate in the eluent. Interesting is the work reported recently by Zhang et al. [86] to produce acetate from CO_2 by using a graphene oxide-coated copper foam electrode (Fig. 11.20) to deposit, as thin biofilm, *Sporomusa ovata* where an acetate production rate of 1.7 mol (100 g) m^{-2} d^{-1} is reported.

The production rate of acetate in MES is not high enough for large-scale application compared with the production of acetic acid from CO_2 and H_2 by anaerobic bacteria [87] of 149 g L^{-1} d^{-1} or with carbohydrates industrial fermentation processes where a productivity of 200 g L^{-1} d^{-1} of acetate is reported [88].

Fig. 11.20 Microbial electrosynthesis reactor with a reduced graphene oxide-coated copper foam composted cathode. Reprinted from Ref. [86], Copyright (2019), with permission from Elsevier

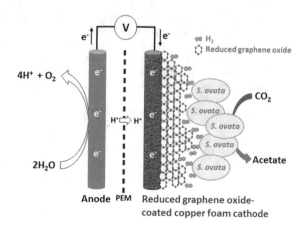

All the above innovative technologies, while presenting new concepts and new routes to complementing Nature in its Carbon-Cycle, need further investigation for rising their TRL to the exploitation level.

References

1. Franceschi VR, Nakata PA (2005) Calcium oxalate in plants: formation and function. Annu Rev Plant Biol 56:41–71
2. Garvie LAJ (2006) Decay of cacti and carbon cycling. Naturwissenschaften 93:114–118
3. Sun Q, Li J, Finlay RD, Lian B (2019) Oxalotrophic bacterial assemblages in the ectomycorrhizosphere of forest trees and their effects on oxalate degradation and carbon fixation potential. Chem Geol 514:54–64
4. https://inhabitat.com/brilliant-cement-making-technology-mimics-coral-while-removing-co2-from-the-atmosphere/
5. https://biomimicry.org/message-to-cop21-leaders/
6. Aresta M, Karimi I, Kawi S (2019) An economy based on CO_2 and water. Springer
7. (a) Dibenedetto A, Colucci A, Aresta M (2016) The need to implement an efficient biomass fractionation and full utilization based on the concept of "biorefinery" for a viable economic utilization of microalgae. Env Sci Pollut Res 23:22274–22283; (b) Dibenedetto A, Aresta M, Dumeignil F (2015) Biorefineries: an introduction. De Gruyter. ISBN 978-3-11-033153-0
8. Thurman HV (1997) Introductory oceanography. Prentice Hall College, New Jersey, USA
9. Adesanya VO, Cadena E, Scott SA, Smith AG (2014) Life cycle assessment on microalgal biodiesel production using a hybrid cultivation system. Bioresour Technol 163:343–355
10. Aresta M, Dibenedetto A, Barberio G (2005) Utilization of macro-algae for enhanced CO_2 fixation and biofuels production: development of a computing software for an LCA study. Fuel Process Technol 86(14-15):1679–1693
11. Jez S, Spinelli D, Fierro A, Dibenedetto A, Aresta M, Busi E, Basosi R (2017) Comparative life cycle assessment study on environmental impact of oil production from micro-algae and terrestrial oilseed crops. Bioresour Technol 239:266–275
12. Dibenedetto A (2019) Enhanced fixation of CO_2 in land and aquatic biomass. In: Aresta M, Kawi S, Karimi IA (eds) An economy based on carbon dioxide and water. Springer. ISBN 978-3-030-15868-2 (chapter 9)
13. https://www.aquaculturealliance.org/advocate/farming-algal-fuel-economics-challenge-process-potential/
14. (a) Harvesting and drying of algal biomass: https://link.Springer.com/chapter10.1007/978-981-13-2378-2_5; (b) A new reactor design fo0r harvesting algae through electrocoagulation-flotationin a continuous mode_Science Direct. https://www.sciencedirect.com/science/article/pii/S2219264419310926
15. de Farias Silva CE, Sforza E, Bertucco A (2019) Enhancing carbohydrate productivity in photosynthetic microorganism production: a comparison between cyanobacteria and microalgae and the effect of cultivation systems. In: Advances in feedstock conversion technologies for alternative fuels and bioproducts. Woodhead Publishing, pp 37–67
16. Rawat I, Kumar RR, Mutanda T, Bux F (2013) Biodiesel from microalgae: a critical evaluation from laboratory to large scale production. Appl Energy 103:444–467
17. Dibenedetto A, Aresta M (2019) Beyond fractionation in the utilization of microalgal components. In: Pires JCM, da Cunha Goncalves AL (eds) Bioenergy with carbon capture and storage. Elsevier. ISBN 9780128162293 (chapter 9)
18. Aresta M, Cornacchia D, Dibenedetto A (2017) Polyfunctional mixed oxides for the oxidative cleavage of lipids and methyl ester of fatty acids. WO2017202955A1

19. Werpy T, Petersen G (2004) Top value added chemicals from biomass volume I—results of screening for potential candidates from sugars and synthesis gas. Technical report NREL/TP-510-35523

20. Holladay JE, White JF, Bozel JJ, Johnson D (2007) Top value-added chemicals from biomass: volume II-results of screening for potential candidates from biorefinery lignin. Technical report PNNL-16983

21. Aresta M, Dibenedetto A (2018) Fuels from recycled carbon. In: Gude VG (ed) Green chemistry for sustainable biofuel production. Apple Academic Press, pp 79–152. ISBN 9781771886390 (chapter 2)

22. Brown RC (ed) (2019) Thermochemical processing of biomass: conversion into fuels, chemicals and power. Wiley

23. Sharma HK, Xu C, Qin W (2019) Biological pretreatment of lignocellulosic biomass for biofuels and bioproducts: an overview. Waste Biomass Valorization 10(2):235–251

24. Di Donato P, Finore I, Poli A, Nicolaus B, Lama L (2019) The production of second generation bioethanol: the biotechnology potential of thermophilic bacteria. J Cleaner Prod 233:1410–1417

25. Hu L, Wu Z, Xu J, Sun Y, Lin L, Liu S (2014) Zeolite-promoted transformation of glucose into 5-hydroxymethylfurfural in ionic liquid. Chem Eng J 244:137–144

26. Abou-Yousef H, Hassan EB (2014) A novel approach to enhance the activity of H-form zeolite catalyst for production of hydroxymethylfurfural from cellulose. J Ind Eng Chem 20:1952–1957

27. Ordomsky VV, van der Schaaf J, Schouten JC, Nijhuis TA (2012) The effect of solvent addition on fructose dehydration to 5-hydroxymethylfurfural in biphasic system over zeolites. J Catal 287:68–75

28. Xu H, Miao Z, Zhao H, Yang J, Zhao J, Song H, Liang N, Chou L (2015) Dehydration of fructose into 5-hydroxymethylfurfural by high stable ordered mesoporous zirconium phosphate. Fuel 145:234–240

29. Jimenez-Morales I, Teckchandani-Ortiz A, Santamaria-Gonzalez J, Maireless-Torres P, Jimenez-Lûpez A (2014) elective dehydration of glucose to 5-hydroxymethylfurfural on acidic mesoporous tantalum phosphate. Appl Catal B 144:22–28

30. Ordomsky V, Van der Schaaf J, Schouten JC, Nijhuis TA (2013) Glucose dehydration to 5-hydroxymethylfurfural in a biphasic system over solid acid foams ChemSusChem 6:1697–1707

31. Zhang Y, Wang J, Ren J, Liu X, Li X, Xia Y, Lu G, Wang Y (2012) Mesoporous niobium phosphate: an excellent solid acid for the dehydration of fructose to 5-hydroxymethylfurfural in water. Catal Sci Technol 2:2485–2491

32. Zhuang J, Lin L, Pang C, Liu Y (2011) Selective catalytic conversion of glucose to 5-hydroxymethylfurfural over Zr (H$_2$PO$_4$)$_2$ solid acid catalysts. Adv Mater Res 236–238:134–137

33. Dibenedetto A, Aresta M, Pastore C, di Bitonto L, Angelini A, Quaranta E (2015) Conversion of fructose into 5-HMF: a study on the behaviour of heterogeneous cerium-based catalysts and their stability in aqueous media under mild conditions. RSC Adv 5:26941–26948

34. Dibenedetto A, Aresta M, di Bitonto L, Pastore C (2016) Organic carbonates: efficient extraction solvents for the synthesis of HMF in aqueous media with cerium phosphates as catalysts. Chemsuschem 9(1):118–125

35. Kläusli T, Biochem AVA (2014) Commercialising renewable platform chemical 5-HMF. Green Process Synth 3:235–236

36. Executive Summary EU Project BIOCONSEPTS, Specialty and Bio Based Polymers, RND_001346

37. Ventura M, Aresta M, Dibenedetto A (2016) Selective aerobic oxidation of 5-(Hydroxymethyl)furfural to 5-formyl-2-furancarboxylic acid in water. Chemsuschem 9(10):1096–1100

38. Dibenedetto A, Ventura M, Lobefaro F, de Giglio E, Distaso M, Nocito F (2018) Selective aerobic oxidation of 5-(hydroxymethyl) furfural to 2,5-diformylfuran or 2-formyl-5-furancarboxylic acid in water using MgO· CeO2 mixed oxides as catalysts. Chemsuschem 11(8):1305–1315

39. Dibenedetto A, Ventura M, Lobefaro F, de Giglio E, Altomare A, Cometa S, Nocito F (2018) Tunable mixed oxides based on CeO_2 for the selective aerobic oxidation of 5-(hydroxymethyl) furfural to FDCA in water. Green Chem 20(17):3921–3926

40. Dibenedetto A, Ventura M, Williamson D, Lobefaro F, Jones MD, Mattia D, Nocito F, Aresta M (2018) Sustainable synthesis of oxalic (and succinic) acid via aerobic oxidation of C6 polyols by using M@ CNT/NCNT (M = Fe, V) based catalysts in mild conditions. Chemsuschem 11(6):1073–1081

41. Omoruyi U, Page S, Hallett J, Miller PW (2016) Homogeneous catalyzed reactions of levulinic acid: to γ-valerolactone and beyond. Chemsuschem 9(16):2037–2047

42. Dutta S, Iris KM, Tsang DC, Ng YH, Ok YS, Sherwood J, Clark JH (2019) Green synthesis of gamma-valerolactone (GVL) through hydrogenation of biomass-derived levulinic acid using non-noble metal catalysts: a critical review. Chem Eng J 372:992–1006

43. Anbarasan P, Baer ZC, Sreekumar S, Gross E, Binder JB, Blanch HW, Clark DS, Toste FD (2012) Integration of chemical catalysis with extractive fermentation to produce fuels. Nature 491:235–239

44. Hilpmann G, Steudler S, Ayubi MM, Pospiech A, Walther T, Bley T, Lange R (2019) Combining chemical and biological catalysis for the conversion of hemicelluloses: hydrolytic hydrogenation of xylan to xylitol. Catal Lett 149(1):69–76

45. Aresta M, Dibenedetto A (2007) Utilisation of CO_2 as a chemical feedstock: opportunities and challenges. Dalton Trans 2975

46. Aresta M, Dibenedetto A, Angelini A (2013) The changing paradigm in CO_2 utilization. J CO_2 Utiliz 3:65–73

47. Subashchandrabose SR, Ramakrishnan B, Megharaj M, Venkateswarlu K, Naidu R (2013) Mixotrophic cyanobacteria and microalgae as distinctive biological agents for organic pollutant degradation. Environ Int 51:59–72

48. Yelton AP, Acinas SG, Sunagawa S, Bork P, Pedrós-Alió C, Chisholm SW (2016) Global genetic capacity for mixotrophy in marine picocyanobacterial. ISME J 10:2946–2957

49. https://www.marketsandmarkets.com/Market-Reports/pha-market-395.html

50. https://www.transparencymarketresearch.com/2-3-butanediol-market.html

51. Han MW, Hyun JL (2017) Toward solar biodiesel production from CO_2 using engineered cyanobacteria. FEMS Microbiol Lett 364:9

52. Gao Z, Zhao H, Li Z, Tan X, Lu X (2012) Photosynthetic production of ethanol from carbon dioxide in genetically engineered cyanobacteria. Energy Environ Sci 5:9857–9865

53. Angermayr SA, van der Woude AD, Correddu D (2014) Exploring metabolic engineering design principles for the photosynthetic production of lactic acid by *Synechocystis* sp. PCC6803. Biotechnol Biofuels 7:99

54. Wang Y, Sun T, Gao X (2016) Biosynthesis of platform chemical 3-hydroxypropionic acid (3-HP) directly from CO_2 in cyanobacterium *Synechocystis* sp. PCC 6803. Metab Eng 34:60–70

55. Kusakabe T, Tatsuke T, Tsuruno K (2013) Engineering a synthetic pathway in cyanobacteria for isopropanol production directly from carbon dioxide and light. Metab Eng 20:101–108

56. Hirokawa Y, Maki Y, Tatsuke T (2016) Cyanobacterial production of 1,3-propanediol directly from carbon dioxide using a synthetic metabolic pathway. Metab Eng 34:97–103

57. Lan EI, Ro SY, Liao JC (2013) Oxygen-tolerant coenzyme A-acylating aldehyde dehydrogenase facilitates efficient photosynthetic n-butanol biosynthesis in cyanobacteria. Energy Environ Sci 6:2672–2681

58. Shen CR, Liao JC (2012) Photosynthetic production of 2-methyl-1-butanol from CO_2 in cyanobacterium *Synechococcus elongatus* PCC7942 and characterization of the native acetohydroxyacid synthase. Energy Environ Sci 5:9574–9583

59. Kanno M, Carroll AL, Atsumi S (2017) Global metabolic rewiring for improved CO_2 fixation and chemical production in cyanobacteria. Nat Commun 8:14724

60. Atsumi S, Higashide W, Liao JC (2009) Direct photosynthetic recycling of carbon dioxide to isobutyraldehyde. Nat Biotech 27:1177–1180

61. https://www.photanol.com/

62. https://phytonix.com/

63. https://www.novamont.com/eng/read-press-release/mater-biotech/

64. https://www.grandviewresearch.com/industry-analysis/1-4-butanediol-market

65. https://bioenergyinternational.com/research-development/30229

66. Kondaveeti S, Kakarla R, Kim HS, Kim BG, Min B (2017) The performance and long-term stability of low-cost separators in single-chamber bottle-type microbial fuel cells. Environ Technol 1–10

67. Rozendal RA, Hamelers HV, Rabaey K, Keller J, Buisman CJ (2008) Towards practical implementation of bioelectrochemical wastewater treatment. Trends Biotechnol 26(8):450–459

68. Lovley DR, Nevin KP (2013) Electrobiocommodities: powering microbial production of fuels and commodity chemicals from carbon dioixde with electricity. Curr Opin Biotechnol 24:385–390

69. Blankenship RE, Tiede DM, Barber J, Brudvig GW, Fleming G, Ghirardi M, Gunner MR, Junge W, Kramer DM, Melis A, Moore TA, Moser CC, Nocera DG, Nozik AJ, Ort DR, Parson WW, Prince RC, Sayre RT (2011) Comparing photosynthetic and photovoltaic efficiencies and recognizing the potential for improvement. Science 332:805–809

70. Chiranjeevi P, Bulut M, Breugelmans T, Patil SA, Pant D (2019) Current trends in enzymatic electrosynthesis for CO_2 reduction. Curr Opin Green Sustain Chem 16:65–70

71. Huang YX, Hu Z (2018) An integrated electrochemical and biochemical system for sequential reduction of CO_2 to methane. Fuel 220:8–13

72. Villano M, Aulenta F, Ciucci C, Ferri T, Giuliano A, Majone M (2010) Bioelectrochemical reduction of CO_2 to CH_4 via direct and indirect extracellular electron transfer by a hydrogenophilic methanogenic culture. Bioresour Technol 101(9):3085–3090

73. Feng Q, Song YC, Ahn Y (2018) Electroactive microorganisms in bulk solution contribute significantly to methane production in bioelectrochemical anaerobic reactor. Bioresour Technol 259:119–127

74. Liu D, Roca-Puigros M, Geppert F, Caizan-Juanarena L, Na Ayudthaya SP, Buisman C, ter Heijne A (2018) Granular carbon-based electrodes as cathodes in methane-producing bioelectrochemical systems. Front Bioeng Biotechnol 6:1–10

75. Zhen G, Zheng S, Lu X, Zhu X, Mei J, Kobayashi T, Xu K, Li YY, Zhao Y (2018) A comprehensive comparison of five different carbon-based cathode materials in CO2 electromethanogenesis: long-term performance, cell-electrode contact behaviors and extracellular electron transfer pathways. Bioresour Technol 266:382–388

76. Zeppilli M, Chouchane H, Scardigno L, Mahjoubi M, Gacitua M, Askri R, Majone M (2020) Bioelectrochemical vs hydrogenophilic approach for CO_2 reduction into methane and acetate. Chem Eng J 125243

77. Yuan M, Kummer MJ, Minteer SD (2019) Strategies for bioelectrochemical CO_2 reduction. Chem Eur J 25(63):14258–14266

78. Batlle-Vilanova P, Rovira-Alsina L, Puig S, Dolors Balaguer M, Icaran P, Monsalvo VM, Rogalla F, Colprim J (2019) Biogas upgrading, CO_2 valorisation and economic revaluation of bioelectrochemical systems through anodic chlorine production in the framework of wastewater treatment plants. Sci Total Environ 690:352–360

79. De Godos I, Cano R, Santiago JR, Lara E, Llamas B (2015) Device and method for simultaneous removal of hydrogen sulphide and carbon dioxide from biogas. EP 3061515 A1

80. Batlle-Vilanova P, Puig S, Gonzalez-Olmos R, Balaguer MD, Colprim J (2016) Continuous acetate production through microbial electrosynthesis from CO_2 with microbial mixed culture. J Chem Technol Biotechnol 91(4):921–927

81. Pavel M, Pant D, Patra S (2017) Integrated photobioelectrochemical systems: a paradigm shift in artificial photosynthesis. Trends Biotechnol 35(4):285–287

82. Aresta M, Dibenedetto A, Baran T, Angelini A, Łabuz P, Macyk W (2014) An integrated photocatalytic-enzymatic system for the reduction of CO_2 to methanol in bio-glycerol-water. Beilstein J Org Chem 10:2556–2565

83. Aresta M, Dibenedetto A, Macyk W, Baran T (2013) Photocatalysts working in the visible region for the reduction of NAD^+ to NADH within an hybrid chemo-enzymatic process of CO_2 reduction to methanol. MI2013A001135

84. Schlager S, Dibenedetto A, Aresta M, Apaydin DH, Dumitru LM, Neugebauer H, Sariciftci NS (2017) Biocatalytic and bioelectrocatalytic approaches for the reduction of carbon dioxide using enzymes. Energy Technol 5(6):812–821

85. Bajracharya S, van den Burg B, Vanbroekhoven K, De Wever H, Buisman CJ, Pant D, Strik DP (2017) In situ acetate separation in microbial electrosynthesis from CO2 using ion-exchange resin. Electrochim Acta 237:267–275

86. Aryal N, Wan L, Overgaard MH, Stoot AC, Chen Y, Tremblay PL, Zhang T (2019) Increased carbon dioxide reduction to acetate in a microbial electrosynthesis reactor with a reduced graphene oxide-coated copper foam composite cathode. Bioelectrochemistry 128:83–93

87. Morinaga T, Kawada N (1990) The production of acetic acid from carbon dioxide and hydrogen by an anaerobic bacterium. J Biotechnol 14(2):187–194

88. López-Garzón CS, Straathof AJJ (2014) Recovery of carboxylic acids produced by fermentation. Biotechnol Adv 32:873–904

The CO$_2$ Revolution

12

Abstract

The potential of carbon recycling through CO$_2$ utilization is "sized" here by merging all options discussed in previous chapters. It is interesting to note that in the medium-long term some 7–9 Gt$_{CO2}$/y can be avoided, and this is a very interesting news: coupled to the use of biomass and utilization of perennial energies (SWHG) will make possible the reduction of the extraction of fossil fuels to, most likely, one-third–one-fourth with respect to today. This will be a great achievement.

12.1 Is CO$_2$ Majorly Responsible for Climate Change?

The emission of CO$_2$ from fossil-C is apparently revolutionizing our planet and our life through climate change. Such tiny molecule is accused to be responsible of potential monstrous disasters. The correlation temperature-atmospheric level of CO$_2$ is under deep debate; some reports say that the trend is parallel, others show that they are antiparallel. Figure 12.1 is, particularly, liked by people who deny the correlation between climate change and CO$_2$ atmospheric level [2a]. Other scientists like more the trend of temperature and CO$_2$ atmospheric level represented in Fig. 12.2 [2b].

Where is the reality? Since long time we have risen the question: is CO$_2$ really responsible of Climate Change [1, 3]? Let us see what is known and the arguments for a tentative answer to such a question. As we have already seen in Chaps. 1–3, the emission of CO$_2$ is a witness of the amount of fossil-C (coal, oil, gas) burned, and its accumulation in the atmosphere (from 275 ppm of the pre-industrial age to actual 410+ ppm) witnesses the inability of natural cycles to buffer its emission with a faster uptake by oceans and fixation by plants. In previous chapters, we have also seen that the use we make of chemical energy stored in fossil-C is very

© Springer Nature Switzerland AG 2021
M. Aresta and A. Dibenedetto, *The Carbon Dioxide Revolution*,
https://doi.org/10.1007/978-3-030-59061-1_12

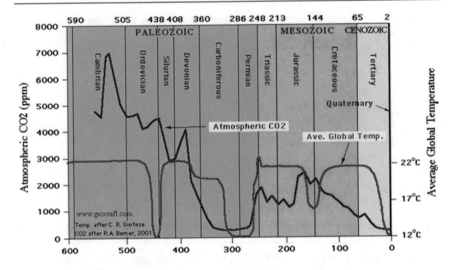

Fig. 12.1 Trend of temperature and CO$_2$ atmospheric level

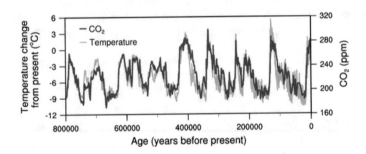

Fig. 12.2 Trend of temperature and CO$_2$ atmospheric level (NOAA data)

inefficient and more energy is transferred to the atmosphere in form of heat than we really use.

> The **Relative Humidity-RH** is defined as the ratio, at a given temperature, of the water vapor pressure in the atmosphere to the equilibrium (saturation) value at that temperature. When saturation is reached, liquid is formed.

Such heat is causing a faster evaporation of water, which is a powerful greenhouse gas. On a global scale [4], the examination of surface data from over 15 000 weather stations and ships during 1975–2004 has shown that the atmospheric relative humidity increased by 0.5–2% per decade over the central and eastern United

States, India, and western China. In each case, this was associated with an increase in: temperature, low clouds, and specific humidity. It looks like atmospheric water vapor level is increasing even at a higher rate than that of CO_2. Therefore, the increase of the concentration of both H_2O and CO_2 is causing trapping IR radiations that cannot escape the atmosphere, increasing the average temperature of the planet. The higher concentration of water vapor causes heavy rains, which we see these days, even in geographical areas which were not used to such phenomena, causing serious disasters, and the reinforcing of such phenomena in equatorial areas. Extreme events are the evidence of climate change that is usually ascribed to the increase of the CO_2 level in the atmosphere, but in our opinion should be more correctly ascribed to the overall inefficient use of fossil-C, which is the real origin of all problems. A question rises quite spontaneous: if we would be able to capture all emitted CO_2 and would continue to warm up the atmosphere at the actual rate by emitting heat, would climate change stop or would we most likely continue to observe an increase of the average temperature of our planet as today, due to the release of heat to the atmosphere and the greenhouse effect of water vapor that is much more abundant and more powerful (see Fig. 3.7) than CO_2? What is, thus, the real contribution of CO_2 atmospheric level to climate change? This question is at the center of debates today and there are several researchers who even deny that CO_2 is responsible for climate change.

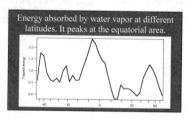

The sophisticated instruments and techniques developed in the last decades, the possibility of exploring the entire atmosphere with local inspection using satellites, have allowed NASA to collect data, not available until recently, which substantiate a more clear vision of the "Climate Change" and who is responsible for what. Dessler says, "We now think the water vapor feedback is extraordinarily strong, capable of doubling the warming due to carbon dioxide alone." Because the new precise observations and measures in situ may support models about the water vapor's impact, researchers can develop more reliable forecast of the trend of our planet temperature. "This study confirms that what was predicted by the models is really happening in the atmosphere." sais Eric Fetzer, an atmospheric scientist who works with AIRS data at NASA's Jet Propulsion Laboratory in Pasadena, Calif. "Water vapor is the big player in the atmosphere as far as climate is concerned." [5].

However, CO_2 is not the protagonist in the Climate Change drama, it is neither the deuteragonist, it is the third actor, with the key positions in this drama we are living taken by *"Protagonist"*: our inability to convert efficiently the chemical energy of fossil-C into other forms of energy; *"Deuteragonist"*: water vapor produced by direct heating of the atmosphere and Earth surface; *"Third actor"*: CO_2.

As we cannot use efficiently fossil-C, biomass alone cannot satisfy our needs, and liquid fuels are the most concentrated forms of energy, our strategy is to substitute fossil-C with renewable carbon, which means with a form of carbon that can enter into a man-made cycle, supplementing the *"natural cycle."*

And CO_2 *"is"* renewable carbon: Nature shows that it is!

12.2 The Carbon Dioxide Revolution

Therefore, the solution is in the *"Carbon Dioxide Revolution"*: turn around the concept about it; *from origin of problems to solution of problems. CO_2 is not a waste*: we are supporting this concept since the 1980s [6]. *CO_2 is a resource*! We have anticipated too much the times! But we are there! People now believe in this paradigm, even industries.

In 1989, we anticipated the use we have to do of CO_2 (Fig. 12.3); after separation, CO_2 can be converted into useful products with the use of solar light and the concourse of catalysis, photochemistry, photoelectrochemistry, and biotechnology; we are there, exactly there, thirty years after. But in the 1980s, there were not yet the conditions for using CO_2 on a large scale.

It is now clear that *Carbon Dioxide Utilization (CDU)* is a step toward a *Cyclic Carbon Economy*, with different options spanned on a diverse time-scale. *CDU represents the revolution in the Society.*

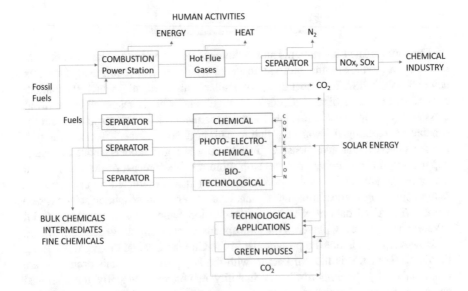

Fig. 12.3 Separation and use of CO_2 [18]

CDU can contribute to the reduction of CO_2 accumulation in the atmosphere through the *"industrial innovation"* that will require less use of natural fossil resources (raw materials diversification), and will be based on more efficient processes with recycling of carbon, implementing the *"industry clustering"* concept.

Moving apart from the fossil-C based energy frame, the big change will require the extended use of largely available and free perennial energy sources. SWGH exploitation, coupled with CO_2 conversion, will bring in the system several advantages in terms of use of existing infrastructures and logistics, with economic benefits. The direct injection of PV- or wind-generated electrons into the electric-grid avoids the penalty (*ca.* 30% of energy loss) typical of the use of electricity for producing hydrogen for the chemical conversion of CO_2 into chemicals. This is a point that many people take in favor of the direct use of electricity. But electricity cannot solve all problems, in the short–medium term. Let us use hydrogen, then! This is the second point against CO_2 conversion, but the direct use of hydrogen rises several problems, such as storage, transportation, change of vehicle engines, safety, and new infrastructure. They will demand a tremendous economic effort and large investments, which will not be affordable in some areas of the planet. Especially safety issues require much attention. The direct use of PV-electricity will be possible for some applications, especially in low-density operation. Some sectors such as avio-transport, navy, and heavy-road transport will not be able to run on PV in the short–medium term and liquid (gaseous) fuels will be necessary for a long time. However, the use of non-fossil H_2 for converting CO_2 into fuels (liquid and gaseous) has a sense from the economic and safety point of view and for guaranteeing the continued utilization of existing and improved infrastructures for the benefit of less advanced countries. As we have already said, intensive energy applications can be only partially de-fossilized and human life cannot be decarbonized. Electricity production for Industries and megalo-polies and heavy-transport will be based on fossil-C derivatives (diesel), which are the highest density energy carriers, still close to the end of this century. Human life will need C-based goods and food forever.

There is not a single technology that can solve the CO_2 problem. Such complex problem needs an integrated solution and CDU is part of the solution package. How much carbon dioxide can, thus, be used and avoided, suppose that all options described in previous chapters are exploited at their best? We recall that *"used"* and *"avoided"* are not synonyms. The former term refers to the amount of CO_2 used in a synthesis (dictated by the stoichiometry), the latter to the amount of CO_2 not emitted (with respect to existing conventional processes) while using CO_2 in syntheses. In an average case, for an "innovative" procedure based on CO_2 carried out in most effective conditions, the ratio "avoided/used" ranges around 2.8. This means that per each tonne of used CO_2, 2.8 are avoided.

We make now an attempt to size the amount of CO_2 that will be used per each class of compounds. Table 12.1 presents the actual and perspective (2040) use of CO_2 in the synthesis of some classes of chemicals, summarizing what has been presented in Chap. 9.

Table 12.1 Uses of CO_2 in the synthesis of chemicals (>1 Mt/y)

Compound	Formula $C_{oxidation\,state}$ (basic unit for polymers)	Market 2019 Mt/y	CO_2 Used Mt/y	Market 2040 Mt/y	CO_2 used Mt/y
Urea	$(H_2N)_2CO$ +4	180	132	366	**266**
Carbonates linear	$OC(OR)_2$ +4	>2	<0.1	10	5
Carbonates cyclic	+4		<0.1		
Polycarbonates	$-[OC(O)OOCH_2CHR]-n$ +4	5	<0.1	9-10	2-3
Polyurethanes	$-[R(H)N-CO_2]-$ +4	ca. 50	<1	>70	*ca. 20*
Poly-acrylates	$CH_2=CHCOOH$ +3	66	<1	>80	40
Formic acid	HCO_2H +2	1	0.9	>10	>9
Inorganic carbonates	M_2CO_3 +4 $CaCO_3$ / M_2CO_3 (Na_2CO_3, K_2CO_3) 250		70	500	**125**
Methanol	CH_3OH -2	85	10	100-200	**130-260**
Total			215		*597-728*
Ethene	$CH_2=CH_2$ -2	160	0	270 (*ca.* 3.5%/y growth)	ca. 170
Propene	$CH_3CH=CH_2$ -2	80	0	150 (*ca.*3.5%/y growth)	ca. 90
Ethene glycol	$HOCH_2CH_2OH$ -1	30	0	50 (*ca.* 5%/y growth)	*ca.* 70
Grand total			207		*927-1 058*

The major consumers of CO_2 are urea, inorganic carbonates, and methanol. The former is by far the largest and oldest application based on CO_2 utilization.

Most of CO_2 used in the urea manufacture is recovered from the synthesis of ammonia, co-reagent with CO_2 in the synthesis of urea, a very old example of industrial CO_2-utilization and C-recycling. The market of urea is expanding for its use in agriculture and for newest applications such as polymers and reactive chemistry. Technologies for the synthesis of polycarbonates are at the demo level and are expanding. Non-isocyanates technologies are required in the synthesis of polyurethanes due to the fact that isocyanates are considered potential cancer agents. Inorganic carbonates can be made through well-known procedures, but using recovered CO_2.

The synthesis of methanol from CO_2 and non-fossil H_2 is at the demo scale, not yet at the industrial full exploitation level, but is a growing interest. On the other hand, it is worth to recall that already today some CO_2 is left in Syngas for the synthesis of methanol since 1970s. The hydrogenation of CO_2 to methane is also at the demo scale [7].

Methodologies for the synthesis of other chemicals are still at a low TRL (3–5), but bottlenecks are known as well as what is needed to boost their exploitation. Noteworthy, the last three chemicals listed in Table 12.1 ethene, propene, and etheneglycol are today totally derived from fossil-C, but routes are known (electrochemical) that may produce such chemicals from CO_2. Assuming that a prudential share of 20% of such chemicals will be produced from CO_2 in 2040, considering a growth rate of their market equal to 3.5% per year [8] an interesting amount of *ca.* 330 Mt/y of used CO_2 can be calculated for the three chemicals.

However, one can foresee that the CO_2 utilization in the synthesis of chemicals that can rise from actual *ca.* 200 Mt to *ca.* 927–1058 Mt by 2040. Assuming a *ratio avoided/used* = 2.8, it can be calculated that 2.6–2.9 Gt of CO_2 can be avoided. The way to assess the benefit of using CO_2 is the LCA methodology [9, 10], which has been more recently used to assess the synthesis of C1 molecules from CO_2 [11]; it has been shown that, among all others, the benefit is maximized for formic acid that has the smaller market (600–800 kt/y). One can argue that such application will have only a small impact on CC: indeed, any contribution of the size of 1 Mt/y can be worth to be considered in making the overall balance.

Microalgae, grown for the production of chemicals and fuels, would greatly contribute to CO_2 fixation (*ca.* 1.8–2.2 t_{CO2} per t of dry biomass) and conversion into fine chemicals (see Chap. 11) and fuels, with significant emission reduction. Products derived from microalgae find application as fine chemicals, food and feed, and fuels [12]; their production from CO_2 will reduce the emission and organic waste that would be generated using classic chemical routes.

Finally, it is the conversion of CO_2 into fuels that have a market 15 times higher than chemicals, which attracts attention as it would be responsible for a major CO_2 conversion and reduction of CO_2 emissions.

It is not realistic to believe that the energy sector will in future be totally based on synthetic fuels. As we have discussed before, fuels derived from CO_2 hydrogenation can be made for some specific transport sectors such as avio, navy, and urban. The CO_2 hydrogenation to fuels will slowly grow until 2040 when the installed PV power will reach >3500 GW installed (with respect to *ca.* 300 today), and the price of PV-H_2 should be comparable to that of MR-H_2, large-scale and durable electrolysers will be available, together with new technologies such as high-pressure electrolysis (for making H_2 under pressure ready for direct utilization in chemical units) and high temperature (solid-state) electrolysis for making hydrogen at the utilization temperature in chemical processes.

Concurrently, catalysts for the selective conversion of CO_2 into liquid fuels (HCn), methanol, and methane should be developed so that all such options will be available allowing large-scale fuels production. The use of Concentrators of Solar Power (CSP) may make use of the simultaneous recovery from the atmosphere of CO_2 and H_2O, with thermal-catalyzed water and CO_2 splitting with production of "Syngas" that would generate "air diesel." Such fuels are much above the market price today.

By integrating catalysis and biotechnology (see Chap. 11) and developing the exploitation of new BES, a consistent bunch of technologies should be available by 2040 for the conversion of several Gt/y of CO_2 into useful products, avoiding consistent amount of fossil-C extraction and use. All such options are summarized in Scheme 12.1.

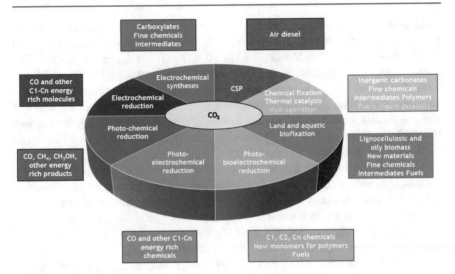

Scheme 12.1 CCU: Technologies and products. Thermal routes to chemicals are in the bluish frame, all other options are based on the use of solar energy either for making electrons or for direct photochemical conversion. *Reprinted/adapted from Ref. [1], Copyright 2019 by permission of Springer*

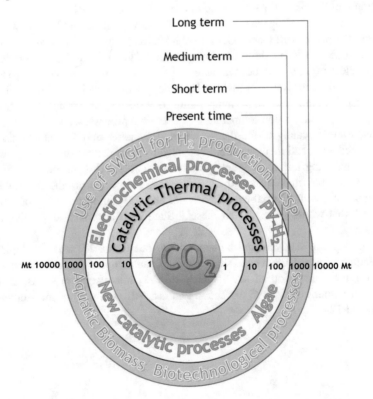

Scheme 12.2 Perspective use of CO₂ that can be converted into chemicals and fuels. Short term: 5–7 y; Medium term: 8–12 y; Long term: 15+ y. *Reprinted/adapted from Ref. [1], Copyright 2019 by permission of Springer*

A study carried out by the Catalyst Group [13] has shown that by combining the potential of all technologies it will be possible by 2040 to convert some 7–9 Gt_{CO2}/y; a figure that agrees with the data from other studies. This is a very interesting target that will contribute to CC control (Scheme 12.2).

As discussed in Chap. 1, the implementation of value-chains and of system clustering of processes may largely improve the utilization of CO_2 and maximize the benefits.

Our personal dream [14–17] is now becoming reality.

The CO₂ Revolution has started

Let us live it taking an active part! It is not only a dream, it is reality.

No land animal can run faster than a car,
no domestic animal can compete with tractors-power,
no fish can swim faster than a boat,
no bird can fly higher and faster than airplanes.
Maybe in the future a device will perform better than plants and microorganisms
in converting CO₂, water, and sunlight into chemical energy.

References

1. (a) Aresta M, Kawi S, Karimi IA (eds) (2019) An economy based on carbon dioxide and water. Springer Publ. ISBN 978-3-030-15868-2; (b) Aresta M (2019), Perspective look on CCU large-scale exploitation. In: Aresta M, Kawi S, Karimi IA (eds) An economy based on carbon dioxide and water, chapter 13. Springer Publ. ISBN 978-3-030-15868-2
2. (a) https://www.discovermagazine.com/environment/heres-what-real-science-says-about-the-role-of-co2-as-earths-preeminent; (b) https://www.ncdc.noaa.gov/global-warming/temperature-change
3. Aresta M, van Eldik R (eds) (2014) Advances in inorganic chemistry, CO_2 chemistry, vol 66, Elsevier Publ
4. Dai A (2006) Recent climatology, variability, and trends in global surface humidity. J Clim 19 (15):3589–3606
5. Dessler AE, Zhang Z, Yang P (2008) Water-vapor climate feedback inferred from climate fluctuations, 2003–2008. Geophys Res Lett 35
6. https://www.nasa.gov/topics/earth/features/vapor_warming.html
7. https://www.caranddriver.com/news/a15370096/audi-opens-first-e-gas-synthetic-fuel-production-facility/
8. https://www.vci.de/langfassungen-pdf/vci-analyis-on-the-future-of-basic-chemicals-production-in-germany.pdf
9. Aresta M, Galatola M (1999) Life cycle analysis applied to the assessment of the environmental impact of alternative synthetic processes. The dimethylcarbonate case: part 1. J Clean Prod 7:181–190
10. Aresta M, Caroppo A, Dibenedetto A, Narracci M (2002) Life cycle assessment (LCA) applied to the synthesis of methanol. Comparison of the use of syngas with the use of CO_2 and dihydrogen produced from renewables. In: Valer M et al (ed) Book on environmental challenges and greenhouse gas control for fossil fuel utilization in the 21st century, 331–348. Kluwer Academic/Plenum Publisher, New York

11. Sternberg A, Jens CM, Bardow A (2017) Life cycle assessment of CO$_2$-based C1-chemicals. Green Chem 19:2244–2259
12. Dibenedetto A, Aresta M (2019) Beyond fractionation in the utilization of microalgal components. In: Pires JCM, da Cunha Goncalves AL (eds) Bioenergy with carbon capture and storage, chapter 9. Elsevier Publ. ISBN 9780128162293
13. Aresta M, Dibenedetto A, He LN (2012) Analysis of demand for captured CO$_2$ and products from CO$_2$ conversion, November, a report exclusively for members of the Carbon Dioxide Capture & Conversion (CO$_2$CC) Program of The Catalyst Group Resources (TCGR)
14. Aresta M, Forti G (eds) (1987) Carbon dioxide as Carbon Source, NATO ASI Series C206. Reidel Publ
15. Aresta M (ed) (2003) Recovery and utilization of carbon dioxide. Kluwer Acad Publ, Dordrecht
16. Aresta M (ed) (2010) Carbon dioxide as chemical feedstock. Wiley VCH
17. Aresta M, Dibenedetto A, Quaranta E (2016) Reaction mechanisms in carbon dioxide conversion. Springer
18. Aresta M, Schloss JV (eds) Enzymatic and model carboxylation and reduction reactions for carbon dioxide utilization. NATO ASI C 314. Kluwer Publ

Appendix A

Electronic Structure of Atoms, Covalent Bonds, and Oxidation Number of an Atom in a Compound

We recall here some basic concepts relevant to the class of reactions named "*redox*," in which the *oxidation state* of some elements is changed. Some reference concepts and rules are categorized below. Only elements frequently encountered in this Book are discussed.

i. The elements are classified on the basis of their "*atomic number*"- Z, that corresponds to the number of protons in the nucleus, which equals the number of electrons in the neutral species. H has 1 proton and 1 electron and atomic number 1, C has $Z = 6$, N has $Z = 7$, O has $Z = 8$.

ii. Electrons are distributed over different energy levels, K, L, M, N…

iii. Each level is divided into sub-levels: K > only one sublevel named s; L > two sublevels, s and p. M has three sublevels: s, p, d. We stop here as we shall deal with atoms with *valence electrons* in levels K and L in the ground state. Sub levels s, p, d are called orbitals (see Fig. A.1): each has a different space geometry. There is a single s orbital with a spherical symmetry, three separated p orbitals (along the three coordinated axes x, y, z), five d orbitals. In general, a level "n" has a number of sub energy levels equal to n^2. Therefore, when $n = 1 \gg 1^2 \gg 1$ only one sub-energy level is possible and this is labeled: "s" or $1s$; when $n = 2 \gg 2^2 \gg 4$ sub energy levels are possible labeled: $2s\ 2p_x\ 2p_y\ 2p_z$; when $n = 3 \gg 3^2 \gg 9$ sub-energy levels are possible labeled: $3s,\ 3p_x 3p_y 3p_z 3d_z 2,\ 3d_{x2\text{-}y2}\ 3d_{xy} 3d_{xz} 3d_{yz}$.

Each orbital can be occupied by two electrons, each with a different *number of spin*, or direction of rotation around its own axis (rotation to the wright or to the left). Therefore, to each energy level can belong $2 \times n^2$ electrons (and as much different elements, each characterized by a different atomic number Z): Level K with $n = 1$ is completed with $2 \times 1^2 = 2$ element (H [$1s^1$], He [$1s^2$]); Level K is said to be the first Period in the Periodic Table (see Appendix J) and is completed with 2 elements. When $n = 2$ (or L) we have the second Period that is completed with $2 \times 2^2 = 8$ elements (Li, Be, B, C, N, O, F, Ne). The third Period is completed with 18 elements ($n = 3 \gg 2 \times$

© Springer Nature Switzerland AG 2021
M. Aresta and A. Dibenedetto, *The Carbon Dioxide Revolution*,
https://doi.org/10.1007/978-3-030-59061-1

Orbitals *s*

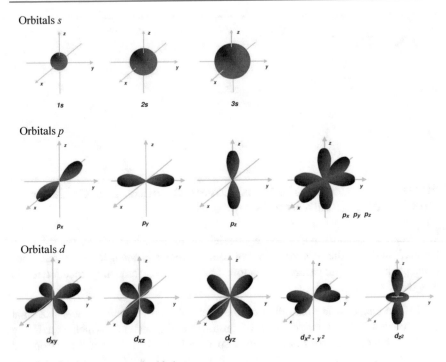

Orbitals *p*

Orbitals *d*

Fig. A.1 Spatial geometry of orbitals

$3^2 = 18$) (see the Periodic Table in the Appendix J for the symbols and electronic structures).

iv. The "*valence electrons*" are those of the uncomplete level. H has one electron in the level K (or 1) and thus has one valence electron. Its electronic structure is indicated as Ks^1 or $1s^1$. He, the second element of the first Period has an electronic structure $1s^2$: the level is complete: the two electrons of He are not considered as valence electrons, He has a scarce reactivity, it is labeled "*noble gas.*" C has six electrons in total, two belonging to the K (or 1) level and four to the L (or 2) level: the latter are "*valence electrons*". Its electronic structure is $K2s^22p^2$. N has an electronic structure $K2s^22p^3$. O has an electronic structure $K2s^22p^4$. Ne ha an electronic structure $K2s^22p^6$ or KL: both are completed and Ne again has no electrons suitable for the formation of bonds at low energy, it is said to be a non-reactive or noble gas, as He. The notation K in C, N, O means that the K level is full, or $1s^2$: such electrons are not valence electrons and will not be implied in bond formation.

v. By convention, all elemental species have oxidation state, n_{ox} equal to zero. So, Fe, Ni, H_2, N_2, O_3, P_4, S_8, have all n_{ox} equal to zero, even if some of them are molecules and not atoms.

vi. A covalent bond between two elements is formed by sharing two "*valence electrons*." "*Valence electrons*" are those that can be implied in bond formation.

vii. Elements can be classified according to their "*attitude to attract electrons*" or "*electronegativity, E*". For the elements we shall deal with, it is $E_H < E_C < E_N < E_O < E_F$.

viii. Between two elements (of the same or different species) can be formed multiple bonds: double bonds such as in O=O, triple bonds such as N≡N or even four bonds.

ix. When a covalent bond is formed, for the calculation of the formal oxidation state of an element, by convention, the bonding pair(s) is(are) attributed to the element that has a higher electronegativity. So, for example, in C=O both electron pairs are attributed to O that is more electronegative than C.

x. The oxidation state of an element is obtained by calculating the difference between the number of *valence electrons (H = 1; C = 4; N = 5; O = 6) and the number of electrons after attributing the bonding electrons to the more electronegative atom*. In covalent compounds, the oxidation state does not represent a net charge on the atoms but is a formal number useful to understand whether the element has been oxidized or reduced in making that specific compound. In general, the highest oxidation state of an element is given by the number of valence electrons, and the lowest oxidation state by the number of lacking electrons for completing the level. In H-H, having the two atoms the same value of E, the two electrons (represented by "-") are attributed one to each atom: therefore each atom has 1 electron as in the atom and the n_{ox} of H in H_2 is zero. In H-F, the electron pair is attributed to F, and the oxidation state of H, is $1 - 0 = +1$. In C = O the four electrons of the double bond (coming two from C and 2 from O) are attributed to oxygen so that C is left with only 2 of the original 4 valence electrons and has an oxidation state of $n_{ox} = 4 - 2 = +2$, while oxygen has $2e^-$ in addition to the original 6 and has an oxidation state of $n_{ox} = 6 - 8 = -2$. In CH_4, H is +1, C is −4; in CH_2O, $n_{oxH} = +1$, $n_{oxO} = -2$, $n_{oxC} = 0$. It is clear that the n_{ox} are just conventional values that may help to understand the electron flow in a reaction.

Note that in a species, the sum of the oxidation states of the elements multiplied each for the relevant stoichiometric coefficient must be equal to the charge of the species. So, in CH_4, being $E_C < E_H$, each H has $n_{ox} = +1$, and C has $n_{ox} = -4$, therefore: $4x(+1)_H - 4_C = 0$ (methane is a neutral species). In CO_3^{2-}, C has $n_{ox} = +4$, O has $n_{ox} = -2$, in total: $+4 + [3x(-2)] = -2$ that is the charge of the ion.

Appendix B

Basics of Thermodynamics

Thermodynamics is a branch of physical-chemistry and physics that deals with temperature (T), Heat (Q), Energy (E), or Work (W). Thermodynamics is based on three classic principles or laws. The principles or laws have been extended to five, with the introduction of the Zero law and Fourth law. Three properties are commonly used in thermodynamics to express the behavior of a reactive system, namely Enthalpy (H) that represents the heat released or up-taken in a process, Entropy (S) that is related to the amount of heat that cannot be transformed into useful work, and Free Energy (G) or Gibbs Energy that gives the useful work made by the system.

The fundamental principles of thermodynamics are categorized below.

The Zero Law

This law states the condition of equilibrium of a system with respect to the environment.

It states:

If two systems are in thermal equilibrium with a third system, then the three systems are in thermal equilibrium with each other.

States of equilibrium are characterized by an *equation of state*, which relates the experimentally accessible parameters, in general: Temperature (T, K), Volume (V, L), Pressure (P, MPa). (The first letter in parentheses is the symbol of the property, the second the unit of measure in the International System of Units.) In a reaction A → B, the equilibrium reaction represents the condition in which the concentrations of the species do not change macroscopically because the rate of the forward process A > B and that of the back process B > A are equal.

© Springer Nature Switzerland AG 2021
M. Aresta and A. Dibenedetto, *The Carbon Dioxide Revolution*,
https://doi.org/10.1007/978-3-030-59061-1

The First Law

The first law establishes that

Energy in a transformation is a "state function", i.e., it only depends on the initial and final
state and not on the pathway used to reach the final state from the origin.

This is expressed by the equation:

$$\Delta E = \Delta W + \Delta Q$$

where E is the internal energy, W is the work done, Q is the heat generated or
dissipated and the symbol Δ indicates the change of the property between the final
and the original state.

The Second Law

The second law is apparently more complex because it is less intuitive. It states that
the entropy of the Universe is a non-decreasing function of time,

$$\Delta S_{Universe} \geq 0$$

The most prominent, and also the oldest expressions of the second law of
thermodynamics are formulated in terms of cyclic processes.
The Kelvin–Planck statement asserts that

no process is possible whose sole result is the extraction of energy from a heat bath, and the
conversion of all that energy into work.

The Clausius statement says

no process is possible whose sole result is the transfer of heat from a body of lower
temperature to a body of higher temperature.

Finally, the Carnot statement declares that

no engine operating between two heat reservoirs can be more efficient than a Carnot engine
operating between those same reservoirs.

The second law is expressed by the equation:

$$\Delta S = \Delta Q/T$$

where S is the Entropy of the system, Q is the amount of Heat exchanged at the
temperature T.

The Third Law

The third law is based on the Nernst theorem that states that as absolute zero of the temperature is approached, the entropy change ΔS for a chemical or physical transformation approaches 0,

$$\lim_{T \to 0} \Delta S = 0$$

This makes that the property S is used to express the disorder or randomness of a system.

The Fourth Law

The fourth law is based on the Onsager reciprocal relations which express the equality of certain ratios between flows and forces in thermodynamic systems out of equilibrium, but where a notion of local equilibrium exists. In absence of magnetic fields, Onsager established

$$L_{j,k} = L_{k,j}$$

that, to a certain degree, makes equation $L_{j,k} = L_{k,j}$ a thermodynamic equivalent of Newton's third law.

A Fundamental Equation in Thermodynamics

A most used equation in thermodynamics is:

$$\Delta G = \Delta H - T\Delta S$$

That correlates the Free Gibbs Energy G (or useful work of a system) to Enthalpy (E) and Entropy (S). The value of the properties at the standard state (T = 273 K, P = 0.101 MPa) is expressed as G°, H°, S°.

Appendix C

Fuel Cells

A fuel cell is a device that converts chemical potential energy (energy stored in chemical covalent bonds) into electrical energy. Such device works on a reduction and an oxidation process which occur at two separated electrodes so that electrons can flow in an external circuit and be used by a utilizer. The two electrode compartments are separated by a protonic membrane that allows H^+ migration while keeping the solution at the same electric potential.

A typical fuel cell is based on the combustion of H_2 with O_2.

In a direct combustion in a thermal engine, the process is represented as

$$H_2 + 1/2\ O_2 = H_2O + \text{Heat.}$$

Therefore, water is formed and heat is released.

If the reaction is carried out at two electrodes in a cell the redox processes are separated:

$$H_2 = 2H^+ + 2e^- \text{(oxidation)}$$

$$1/2O_2 + 2e^- = O^{2-} \text{(reduction)}$$

In this case energy is, thus, released as electric energy instead of heat.

The two compartments of the cell are separated by a Proton Exchange Membrane (PEM) that allows the transfer of protons with formation of water from proton and oxide. The products of the reaction in the cell are water, electricity, and some heat which is lost. The lost heat represents the loss in efficiency in the

© Springer Nature Switzerland AG 2021
M. Aresta and A. Dibenedetto, *The Carbon Dioxide Revolution*,
https://doi.org/10.1007/978-3-030-59061-1

Fig. C.1 Scheme of a fuel cell

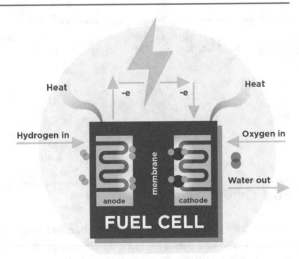

transformation of chemical energy into electricity. This is a big improvement over internal combustion engines which has a much lower efficiency.

Since O_2 is readily available in the atmosphere, it is only necessary to supply the fuel cell with H_2 which can come from a variety of sources.

Other chemicals (e.g., CH_3OH) can be used instead of H_2 (Fig. C.1).

Appendix D

Technology Readiness Level-TRL

TRL is an indicator of the development of a "new technology" (Fig. D.1). It covers the space from Laboratory scale research (TRL = 1–2) to laboratory proof of concept (TRL 3–4), to technology development (TRL = 5–6) and Demo (TRL 7–8) up to on stream processes (TRL = 9).

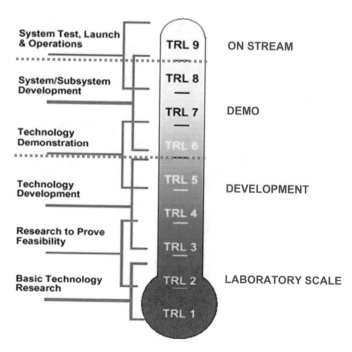

Fig. D.1 The TRL scale

© Springer Nature Switzerland AG 2021
M. Aresta and A. Dibenedetto, *The Carbon Dioxide Revolution*,
https://doi.org/10.1007/978-3-030-59061-1

Appendix E

Table of Hardness: The Mohs Hardness Scale

The Table of Hardness, or Mohs scale, lists 10 materials whose surface presents a different resistance to scratch. Talc is the one that is most easily affected, while diamond is the hardest. A list of different tools, able to scratch different materials, is given in the figure as term of paragon for visualizing the effects (Fig. E.1).

Note: Do not confound hardness and fragility as they are two different properties. Diamond is very hard and very fragile (it will not be scratched by a nail, but will be pulverized by hammer!)

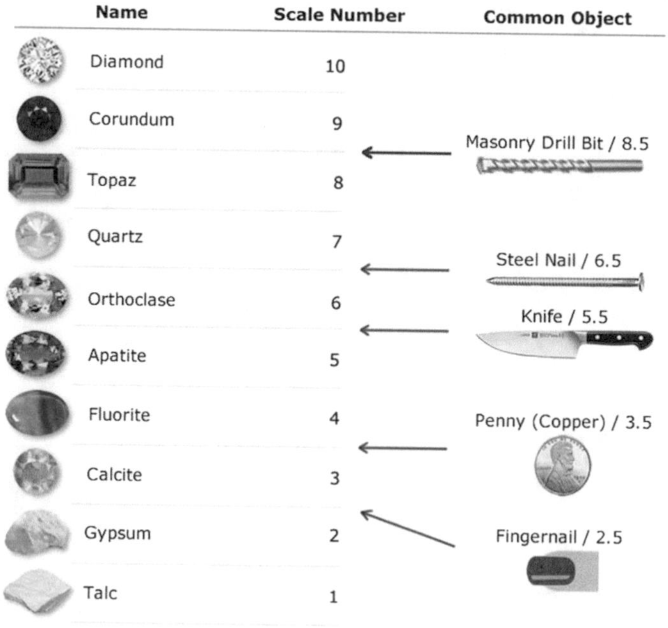

Name	Scale Number	Common Object
Diamond	10	
Corundum	9	
		Masonry Drill Bit / 8.5
Topaz	8	
Quartz	7	
		Steel Nail / 6.5
Orthoclase	6	
		Knife / 5.5
Apatite	5	
Fluorite	4	
		Penny (Copper) / 3.5
Calcite	3	
Gypsum	2	Fingernail / 2.5
Talc	1	

Fig. E.1 Scale of hardness

© Springer Nature Switzerland AG 2021
M. Aresta and A. Dibenedetto, *The Carbon Dioxide Revolution*,
https://doi.org/10.1007/978-3-030-59061-1

Appendix F

See Table E.1.

Table E.1 The prefix symbols of the IS of units

Multiplying factor	SI prefix	Scientific notation
1 000 000 000 000 000 000 000 000	yotta (Y)	10^{24}
1 000 000 000 000 000 000 000	zetta (Z)	10^{21}
1 000 000 000 000 000 000	exa (E)	10^{18}
1 000 000 000 000 000	peta (P)	10^{15}
1 000 000 000 000	tera (T)	10^{12}
1 000 000 000	giga (G)	10^{9}
1 000 000	mega (M)	10^{6}
1 000	kilo (k)	10^{3}
100	hecto (h)	10^{2}
10	deca (da)	10^{1}
1		10^{0}
0.1	deci (d)	10^{-1}
0.01	centi (c)	10^{-2}
0.001	milli (m)	10^{-3}
0.000 001	micro (µ)	10^{-6}
0.000 000 001	nano (n)	10^{-9}
0.000 000 000 001	pico (p)	10^{-12}
0.000 000 000 000 001	femto (f)	10^{-15}
0.000 000 000 000 000 001	atto (a)	10^{-18}
0.000 000 000 000 000 000 001	zepto (z)	10^{-21}
0.000 000 000 000 000 000 000 001	yocto (y)	10^{-24}

Note The *prefix symbol* attaches directly to the symbol for a unit without any spacing: kg, mm, cL

© Springer Nature Switzerland AG 2021
M. Aresta and A. Dibenedetto, *The Carbon Dioxide Revolution*,
https://doi.org/10.1007/978-3-030-59061-1

Appendix G

Life Cycle Assessment and the Impact Categories

Life Cycle Assessment (LCA) is a methodology developed for the comparison, for their environmental impact, of different processes that may exist in the same product. It implies the computation, on the basis of process data (or of best-simulated process data), of energy input-output, in-out mass balance, by-product, and waste production. The impact on a number (13) of categories is evaluated. The expression through a single value of the environmental impact of a given process has also been attempted with alternate fortune as it is not simple to find and "equivalence" among the various "impact categories" and to "weigh" them correctly.

It is a quite complex methodology that requires a careful assessment of data: wrong fed data produce highly erroneous conclusions. The evaluation of the uncertainty of data is of fundamental importance.

The absolute assessment of a process has also been attempted with a high uncertainty about the meaning of conclusions.

Studies carried out by third parties may be considered more reliable than in house carried studies. The latter, on the other hand, can protect the sensitive process data that are not accessible to third parties (Table G.1).

Table G.1 List of environmental impact categories

Impact category/Indicator	Unit	Description
Global warming	kg CO_2-eq	Indicator of potential global warming due to emissions of greenhouse gases to air
Ozone depletion	kg CFC-11-eq	Indicator of emissions to air that cause the destruction of the stratospheric ozone layer
Acidification of soil and water	kg SO_2-eq	Indicator of the potential acidification of soils and water due to the release of gases such as nitrogen oxides and sulphur oxides

(continued)

© Springer Nature Switzerland AG 2021
M. Aresta and A. Dibenedetto, *The Carbon Dioxide Revolution*,
https://doi.org/10.1007/978-3-030-59061-1

Table G.1 (continued)

Impact category/Indicator	Unit	Description
Eutrophication	kg PO_4^{3-}-eq	Indicator of the enrichment of the aquatic ecosystem with nutritional elements, due to the emission of nitrogen or phosphor-containing compounds
Photochemical ozone creation	kg ethene-eq	Indicator of emissions of gases that affect the creation of photochemical ozone in the lower atmosphere (smog) catalyzed by sunlight
Depletion of abiotic resources—elements	kg Sb-eq	Indicator of the depletion of natural non-fossil resources
Depletion of abiotic resources—fossil fuels	MJ	Indicator of the depletion of natural fossil fuel resources
Human toxicity	1,4-DCB-eq	Impact on humans of toxic substances emitted to the environment (Dutch version of EN15804 only)
Fresh water aquatic ecotoxicity	1,4-DCB-eq	Impact on freshwater organisms of toxic substances emitted to the environment (Dutch version of EN15804 only)
Marine aquatic ecotoxicity	1,4-DCB-eq/sup>	Impact on seawater organisms of toxic substances emitted to the environment (Dutch version of EN15804 only)
Terrestrial ecotoxicity	1,4-DCB-eq	Impact on land organisms of toxic substances emitted to the environment (Dutch version of EN15804 only)
Water pollution	m^3	Indicator of the amount of water required to dilute toxic elements emitted into water or soil (french version of EN15804 only)
Air pollution	m^3	Indicator of the amount of air required to dilute toxic elements emitted into air (French version of EN15804 only)

Properties 1–7 should be assessed in any study. Often only the CO_2-footprint of a process is evaluated, that is not enough to correctly determine the environmental impact of a process.

Appendix H

See Table H.1.

Table H.1 Standard Enthalpy and Entropy of formation of some products. The standard Gibbs Free Energy of formation $\Delta G°$ can be calculated by using the $\Delta G° = \Delta H° - T\Delta S°$ equation, where T is the temperature in K

Substance	State	$\Delta H°_f$ $\left(\frac{kJ}{mol}\right)$	$S°$ $\left(\frac{J}{mol\,K}\right)$	Substance	State	$\Delta H°_f$ $\left(\frac{kJ}{mol}\right)$	$S°$ $\left(\frac{J}{mol\,K}\right)$
Ag	s	0	42.6	Cl_2	g	0	223.0
Ag+	aq	105.79	72.7	Cl^-	aq	−167.080	56.5
AgCl	s	−127.01	96.2	ClO_4^-	aq	−128.10	182.0
AgBr	s	−100.4	107.1	Cr	s	0	23.8
$AgNO_3$	s	−124.4	140.9	Cr_2O_3	g	−1139.7	81.2
Al	s	0	28.3	Cu	s	0	33.2
Al^{+3}	aq	−538.4	−321.7	Cu^+	aq	+71.7	40.6
$AlCl_3$	s	−704	110.7	Cu^{+2}	aq	+64.8	−99.6
Al_2O_3	s	−1675.7	50.9	CuO	s	−157.3	42.6
Ba	s	0	62.8	Cu_2O	s	−168.6	93.1
$BaCl_2$	s	−858.6	123.7	CuS	s	−53.1	66.5
$BaCO_3$	s	−1216.3	112.1	Cu_2S	s	−79.5	120.9
$Ba(NO_3)_2$	s	−992	214	$CuSO_4$	s	−771.4	107.6
BaO	s	−553.5	70.4	F^-	aq	−335.35	−13.8
$Ba(OH)_2$	s	−998.2	112	F_2	g	0	202.7
$BaSO_4$	s	−1473.2	132.2	Fe	s	0	27.3
Br_2	l	0	152.2	$Fe(OH)_3$	s	−823.0	106.7
C	s	0	5.7	Fe_2O_3	s	−824.2	87.4
CCl_4	l	−135.4	216.4	Fe_3O_4	s	−1118.4	146.4
$CHCl_3$	l	−134.5	201.7	H_2	g	0	130.6
CH_4	g	−74.8	186.2	H^+	aq	0	0.0ª
C_2H_2	g	+226.7	200.8	HBr	g	−36.29	198.6

(continued)

© Springer Nature Switzerland AG 2021
M. Aresta and A. Dibenedetto, *The Carbon Dioxide Revolution*,
https://doi.org/10.1007/978-3-030-59061-1

Table H.1 (continued)

Substance	State	ΔH°_f $\left(\frac{kJ}{mol}\right)$	S° $\left(\frac{J}{mol\,K}\right)$	Substance	State	ΔH°_f $\left(\frac{kJ}{mol}\right)$	S° $\left(\frac{J}{mol\,K}\right)$
C_2H_4	g	+52.3	219.5	HCO_3^-	aq	−689.93	91.2
C_2H_6	g	−84.7	229.5	HCl	g	−92.31	186.8
C_3H_8	g	−103.8	269.9	HF	g	−273.30	173.7
CH_3OH	l	−238.7	126.8	HI	g	26.50	206.5
C_2H_5OH	l	−277.7	160.7	HNO_3	l	−174.1	155.6
CO	g	−110.53	197.6	HPO_4^{-2}	aq	−1299.0	−33.5
CO_2	g	−393.51	213.6	HSO_4^-	aq	−886.9	131.8
CO_3^{-2}	aq	−675.23	−56.9	H_2O	l	−285.830	69.9
Ca	s	0	41.4	H_2O	g	−241.826	188.7
Ca^{+2}	aq	−543.0	−53.1	$H_2PO_4^-$	aq	−1302.6	90.4
$CaCl_2$	s	−795.8	104.6	H_2S	g	−20.6	205.7
$CaCO_3$	s	−1206.9	92.9	Hg	l	0	76.0
CaO	s	−634.92	39.8	Hg^{+2}	aq	170.21	−32.2
$Ca(OH)_2$	s	−986.1	83.4	HgO	cr, red	−90.79	70.3
$CaSO_4$	s	−1434.1	106.7				
Cd	s	0	51.8				
Cd^{+2}	aq	−75.92	−73.2				
$CdCl_2$	s	−391.5	115.3				
CdO	s	−258.35	54.8				

[a]The standard entropy of the $H^+(aq)$ ion is defined to be 0

Appendix I

Dictionary of acronyms and technical terms

A

Acceptor	An atom that has empty orbitals of suited energy for the formation of a covalent chemical bond with a "donor".
Acid	A species that can transfer protons (Broensted acid) or accept electrons (Lewis acid).
Acid rain	Rain carrying acid species such as H_2SO_4, HNO_3 formed by hydration of acid oxides.
Antennae	Device used to catch solar radiations.
Antisymmetric	A stretching in which two bonds (for example, in triatomic linear molecule ad O=C=O elongate in a opposite mode, one elongates the other shortens.
Avoided	Non-produced: avoided CO_2 is the amount of CO_2 not produced in a process. In an innovative CO_2-based reaction, it represents not only the amount converted but also the CO_2 not emitted with respect to processes on stream.

B

Base	A species that can release OH^- (Broensted base) or share a doublet of electrons with another species (Lewis base) in the formation of a bond.
BES	Bio-electrosystem, the coupling of microorganisms or enzymes with electrodes for an enhanced bioprocess (reduction or oxidation).
Bioeconomy	Exploitation of bioresources (grown or residual biomass).
Biotechnology	A process driven by biosystems, microorganisms or enzymes.
Boudouard reaction	A equilibrium reaction in which C and CO_2 are converted into CO at high temperature.

© Springer Nature Switzerland AG 2021
M. Aresta and A. Dibenedetto, *The Carbon Dioxide Revolution*,
https://doi.org/10.1007/978-3-030-59061-1

Broensted	A Danish chemist who developed the concept of acid and bases. He classified substances according to their ability to either transfer protons to a second species (or even intramolecularly) or to release OH^- moieties in water solution. The former are said *acids* and the latter *bases*.
Building-block	A moiety that enters as such in the molecular structure of a more complex species.
Byproduct	A product formed besides the main target species in a chemical reaction.

C

CAPEX	Capital expenditure, the amount of money necessary to build a plant, investment.
Carbide	Binary compound formed between a metal and carbon, M_xC_y. Among carbides, well known are CaC_2 (that upon reaction with water originates CaO and acetylene, HCCH, used in the past as a lamp gas) and Fe_xC_y, an alloy containing 2–4.3% C used for cast in a mould.
Carbonation	A reaction in which the entire CO_2 moiety is trapped into a product: inorganic or organic carbonate.
Catalyst	A species that can modify the reaction path by modifying the nature of the intermediate and, thus, its energy with respect to the reagents. Usually, a catalyst has a positive effect and, by lowering the energy of the intermediate(s), accelerates the reaction. Anyway, there are also negative catalysts which prevent a reaction to occur.
CCS	Carbon (CO_2) Capture and Storage or disposal in natural fields.
CCU	Carbon (CO_2) Capture and Utilization as a technical fluid or reagent.
Chemical bond	A bond formed between two elements by sharing one or more electron-pairs (covalent bond) or by transfer of one or more electrons (ionic bond).
Chlorofluorocarbons	Molecules in which to a skeleton of bonded C-atoms are linked Cl and F atoms, eventually in addition to H-atoms.
Circular economy	A Nature-like economic system in which waste is not produced.
Climate change	Climatic mutation with respect to historical trend.
Cluster	A number of units (atoms, chemical processes) close in space and interconnected.

Clustering of processes	Building of a variety of interconnected chemical processes in the same site for the optimization of energy and use of resources and minimization of waste.
Coking	Generation of finely divided coke on a catalyst, that causes deactivation.
Conductor	A metal that can conduct electricity: best conductor is silver.
Coordination	The covalent bonding to a metal center of ions (atomic and molecular) or neutral molecules.
CO_2 penalty	The ratio of CO_2 produced to CO_2 fixed in a chemical process. Higher is the value, higher is the penalty.
Covalent bond	A interatomic bond formed by sharing one or more (up to four) electron pairs.
Critical point	A point identified by a value of temperature and pressure above which a gas cannot be liquefied by compression.
Crown-ether	A cyclic ether formed by alternating $(-CHR-)_n$ units (n = 1, 2, 3) and O atoms.
Cumulene	A species that presents two double bonds on the same C-atom, linked to two C-atoms. $H_2C=C=CH_2$, allene, is a cumulene.
Cycle	Turn-around of a species among different physical states (cycle of liquid–solid–vapor water) or a variety of compounds (N_2–NO_x–NH_3) maintaining an overall equilibrium state.

D

DAC	**D**irect **A**ir **C**apture, capture of CO_2 from air where it is present at low concentration (410^+ppm).
Dam	A barrier (usually made of cement, in old days of wood or bricks) constructed to hold back water and raise its level, forming a reserve for either generate electricity or to be used as a water supply.
DMC	Dimethyl carbonate, $(CH_3O)_2CO$ a chemical that finds application in several fields as reagent, solvent or monomer for polymers. DMC is also used as octane number improver.
DME	Dimethyl ether, $(CH_3)_2O$ formed by dehydration of two molecules of methanol, CH_3OH. DME finds use as a substitute of diesel.
Donor	A species that has an electron pair of suitable energy to be used in the formation of a bond with an acceptor.
Dry-Reforming	High-temperature conversion of methane into Syngas by reaction with CO_2 more than with water (*wet-reforming*).

E

Electrophile	A species that has a Lewis acid character and "likes" electrons. It reacts with electron-rich species.
End-on coordination	Co-ordination of a molecule to a metal center through a terminal atom, for example through the O-atom of CO_2 (M < O=C=O).
Eta1-coordination	Coordination of a species through a single, non-terminal atom to a metal center. For example, co-ordination of CO_2 through the only C-atom to a donor.
Electronegativity	The property of a nucleus of attracting electrons, including the bonding electron-pairs.
End of pipe	An operation that occurs at the terminal stage of a process.
Energetic balance	The balance of energy transferred to and released from a reactive process, even very complex and multi-step.
Enthalpy	The thermodynamic property that represents the amount of heat released or up-taken by a reactive process at constant pressure (e.g., occurring in a vessel in contact with the atmosphere). In the former case, the process is said exothermic, in the latter endothermic. Enthalpy symbol is H.
Entropy	The thermodynamic property that indicates the amount of thermal energy of a system that cannot be transformed into useful work. Entropy is also said to represent the randomness or disorder of a system.
EOR	**E**nhanced **O**il **R**ecovery is a technology that allows the recovery of oil from rocks through the injection of water vapor or of supercritical CO_2.
Epoxides	A chemical compound in which one O-atom is linked to two adjacent, bonded C-atoms.
Equilibrium	The state of a chemical reaction represented by equal rate of the forward and back reaction. In such conditions, the macroscopic concentration of the species does not change with time, while the microscopic concentration grows and decreases by the same amount.
EUCheMS	The European Association of National Chemical Societies that promotes the growth and dissemination of Chemical and Materials Science.

F

Fisher-Tropsch	The FT process was developed in Germany by Franz Fischer and Hans Tropsch at the Kaiser-Wilhelm-Institut für Kohlenforschung in Mülheim an der Ruhr, around 1925. It consists of a

number of reactions that convert Syngas (H_2+CO) into hydrocarbons (C_nH_{2n+2}, with n typically in the range 1–20) and some of their isomers and derivatives. Most valuable fractions in a FT-process are gasoline (boiling point 80–120 °C) and diesel (bp 150–400 °C).

Flared-gas Residual process gases (usually formed by light hydrocarbons, CO and hydrogen) which are burned at the exit of a chimney on the industrial site where they are produced.

Footprint The impact on the environment of human activities in terms of pollution, damage to ecosystems, and the depletion of natural resources.

Free energy A thermodynamic property that expresses the amount of energy released (uptaken) in a process that can be converted into useful work.

G

Gibbs Energy Is the "free-energy" of a process that occurs at constant pressure.

Global Capacity Represents the worldwide installed maximum production of a good, chemical, energy, even if not completely used.

Global warming Heating of our Planet that deviates from historical series of data.

Ground state Fundamental energetic state of a system at its equilibrium position.

H

Henry constant A proportionality constant that relates the solubility (equilibrium concentration) of a non-reactive gas in a solvent to the pressure of the gas on the liquid surface.

Heterocumulene A species that presents two double bonds on the same carbon atom, that is not linked to two carbons as in the cumulene: $HN=C=O$, $H_2C=C=O$, $O=C=O$.

Hydrocarbons Gaseous, liquid or solid compounds formed by C and H, of general formula C_nH_{2n+2}. Methane is the first member of the series.

I

IGCC Integrated Gasification Combined Cycle, a new technology that couples C-gasification to Syngas by reaction with H_2O at high temperature, and a heat

	recovery and utilization for producing high temperature and high-pressure water vapor for spinning a turbine and making electric energy.
Initiator	A chemical substance (catalyst) that initiates a reaction that is continued by other drivers.
Incineration bottom ash	Ashes formed in the incineration of waste.
Insulator	A material that does not conduct electric current (electric insulator), heat (thermal insulator), or sound (acoustic insulator).
Ionic Bond	A bond formed by electron transfer from one atom to another with formation of a positively charged species (cation) and a negatively charged one (anion). More electronegative elements will attract preferentially the electrons and will bear the negative charge.
IPCC	International Panel on Climate Change formed by experts designated by different Countries.

J

Joule One of the units of measure of energy. $4.18\ J = 1$ cal.

K

Kinetics Kinetics studies the rate of a reaction (chemical or biochemical) or of a process in general. The rate is determined experimentally.

L

LCA	Life Cycle Assessment, a methodology that compares processes through their impact on a number (usually 13) of categories (see Appendix D).
LCA-E	LCA that includes the economic impact of processes.
LCA-S	LCA that includes the social impact of processes.
Lewis acid or base	*Lewis* acids are those chemical species that have empty orbitals of suitable energy to form a covalent bond using an electron-pair of suited energy of a *Lewis* base.
Ligand	A molecule or charged species (atomic ion or molecular ion) that binds to a metal center through a covalent or ionic bond.
Linear economy	An economic model that uses natural resources for making goods that are disposed after use. Often goods are used only once (single use). This trend causes the formation of large

amounts of waste and reduces the availability of natural resource.

Linear geometry   An arrangement of three or more atoms along a single direction.

M

MAP   Modified Atmosphere Packaging, use of an inert gas instead of air to prevent oxidation.

Mass spectroscopy   Analytical technique that allows to measure the mass of fragments generated from a starting molecular compound by high energy bombardment.

MEA   Monoethanolamine, an organic species containing the amino (NH_2) and alcohol (OH) moieties, of formula $H_2NCH_2CH_2OH$, used for the capture of CO_2.

Merchant market   Global open market of a chemical product.

Mole   Is the counting unit (*dozen*) of Chemists. Its symbol is *mol*. 1 mol corresponds to $6.023.10^{23}$ (the Avogadro number) of something (atoms, molecules, ions, electrons, protons, or eggs).

Molecular orbital   A bonding energy level extended to two or more atoms.

MTBE   Methylterbuthylether, $H_3C–O–C(CH_3)_3$ now used as octane number improver in gasoline instead of lead tetraethyl, $PbEt_4$.

Mtoe   Milliontonequivalent, energy content of a system, of a given amount of fuel expressed as an equivalent amount of oil.

N

NMR   Nuclear Magnetic Resonance, an analytical spectroscopic technique that allows to identify a nucleus and its bounding environment through the specific influence on/by surrounding nuclei in a magnetic field.

Nucleophile   An electron-rich species that reacts with electron-poor species.

n-Type   An extrinsic semiconductor doped with donor species.

O

OECD   Convention on the Organisation for Economic Co-operation and Development that gathers 37 countries, active since 1961. OECD members are Australia, Austria, Belgium, Canada, Chile, Colombia, Czech Republic, Denmark, Estonia, Finland, France, Germany, Greece, Hungary, Iceland, Ireland, Israel, Italy, Japan,

Korea, Latvia, Lithuania, Luxemburg, Mexico, Netherlands, New Zealand, Norway, Poland, Portugal, Slovak Republic, Slovenia, Spain, Sweden, Switzerland, Turkey, United Kingdom, USA.

Olefins — Unsaturated hydrocarbons of general formula C_nH_{2n} containing a C=C double bond, either terminal or internal.

OPEX — OPerational EXpenditure, costs for running a plant or a process. It includes personnel costs, maintenance among others.

Orbital — Energy state of an electron in an atom (atomic orbital) or a molecule (molecular orbital).

Oxidation state — A positive or negative number that indicates the number of electrons a bonded atom in a compound has respectively lost or acquired (formally in covalent bonds or really in ionic bonds) with respect to the elemental state (oxidation state zero).

Ozone hole — Reduction of the stratospheric ozone layer that protects our planet from UV-solar radiations.

P

Pay-back time — Period of time during which investment costs are paid by the net income of a process.

PCC — Pulverized Coal Combustion, for a better and more complete combustion.

PERC — Perchloroethene, $Cl_2C=CCl_2$, a fluid used in dry cleaning.

Perennial — Something that has an endless existence.

Petrochemistry — Chemistry of petrol, processes based on oil conversion.

Phase — A homogeneous system, that has the same properties in all its points.

Photoelectrochemical — A system that generates a flux of electrons in an electrochemical cell upon light irradiation over electrodes made of semi-conducting materials.

Photosensitizer — A material that is excited by light irradiation and can transfer its energy to other species.

Photovoltaic cell — A device that generates a difference of electrical potential and an electrical current under solar irradiation of semiconductors.

Point group — Any of the 32 sets of symmetry operations which can be used to characterize three-dimensional lattices and are the basis of the system of crystal classes.

Polycarbonates — Polymers made of ordered alternated units of an organic moiety and the $-OC(O)O-$ group-of-atoms: $-[CH_2-CH_2-O-C(O)O]_n-$ is ethene carbonate.

Polymers — A material made of a repetitive sequence of one single unit (monomer) or of two co-monomers.

Polyurethanes	A material made of the alternated organic moiety and the $-NH(R)-C(O)O-$ group-of-atoms.
PPB	Part per billion, a unit of concentration of pollutants.
PPM	Part per million, a unit of concentration of pollutants.
Propagator	A center (catalyst) that propagates a reaction.
p-Type	An extrinsic semiconductor doped with electron acceptor atoms.

Q

Quad-generation	Evolution of co-generation biomass-based plant that merges combined power, heat, cooling, and synthetic natural gas production.

R

Radical	A species formed by homolytic cleavage of a covalent bond. Radical anion is formed by transferring a single electron to a neutral species, a radical cation is obtained by taking a single electron from a neutral species.
Raman spectroscopy	A spectroscopic technique used with apolar species inactive in the infrared-IR. By inducing an asymmetry in the species, it becomes IR-active.
Red mud	Mud formed in the treatment of bauxite (Aluminum ore). The red color is caused by iron oxides present in the bauxite.
Reforming	High-temperature treatment of coal or methane for making Syngas, a mixture of H_2 and CO.
Renewable	That can be reformed in time, typical of biomass. Sometimes, such adjective is used also with solar, wind, hydro and geothermal energy, better qualified as perennial energy sources.
Reverse WGS	The reverse of the water gas shift reaction, in which CO_2 is reacted with hydrogen to produce CO and H_2O.

S

Scattering	The process in which electromagnetic radiations or particles are deflected or diffused
Semiconductor	A material with intermediate properties between a metal conductor and an insulator.

Shale fracking	Shale is a soft finely stratified sedimentary rock formed from consolidated mud or clay and can be split easily into fragile plates. Fracking is the process of injecting fluid at high pressure into subterranean rocks (shale) so as to force open existing fissures for oil or gas extraction. Shale-fracking of natural gas is very popular in these days.
Side-on coordination	Co-ordination to a metal center through two adjacent atoms of a ligand: side-on coordination of CO_2 through the C and one of the two O-atoms.
Solar spectrum	Ensemble of radiations emitted by sun.
Solar chemistry	Chemical processes driven by solar energy, opposed to thermal processes powered with fossil-C.
Solar driven	Processes energized with solar radiations.
Solubility	The equilibrium concentration of a non-reactive solute in a solvent.
Sorbent	A material that can absorb a fluid, in general a gas.
Source of energy	Naturally available material or physical agent that can provide energy.
Spin (electron)	Rotation of the electron around its own axis, to the left or right.
Steel slag	Mixture of oxides formed in the production of steel. It may contain basic or acid oxides, and even trace metals which can be toxic.
Stretching	Elongation or shortening of the equilibrium distance typical of two bonded atoms.
Supercritical	State of a gas that does not allow condensation to a liquid.
Sustainable	An ecofriendly process that is durable in time and does not cause depletion of natural resources.
Symmetric	The similar (two elongations or two shortening) stretching of two bonds in a triatomic system.
Syngas	Synthesis gas made of a mixture of H_2 and CO, made from coal, methane, biomass.

T

Thermodynamics	A branch of physical-chemistry that deals with the relationship among heat, temperature, work, and energy. It may be extended to the relationship among various forms of energy.
Triple point	A point, identified by a particular value of T and P in a phase diagram, at which the three phases solid, liquid, gas co-exist. This is a singular point characteristic per each species.

U

Ultramafic   Rocky materials containing **Magnesium** and **Ferric** compounds.
UN-FCCC   United Nations Framework on Climate Change.
Unit-cell   Basic motif that is repeated in the tree spatial dimensions, characteristic of any crystalline species. It may contain one-or-more unit formula of a species such as metal oxides.

V

Valence   The capacity of an atom of making covalent bonds. It is given, in general, by the number of electrons in the uncomplete level of energy. In ionic compounds the valence is given by the charge of the cation or anion.
Value-chain   A chain of processes that can convert a raw material or a waste in a sequence of added-value products with multiple use.
Vector of energy A non-natural resource/material that can produce energy typically by combustion (e.g., H_2).
Visible light   Light seen by human eyes.

W

Water-Gas-Shift A process that maximizes the formation of H_2 from coal or methane in their reaction with water. Typically, it is based on the further reaction of CO with H_2O to afford H_2 and CO_2.
Weathering   Natural process in which a basic natural silicate is converted into a carbonate and silica by reaction with atmospheric CO_2 and water.
Wet-Reforming   A process in which methane is reacted with water at high temperature to afford Syngas (H_2 + CO).

Z

Zero emission   A process in which in which energy is generated without net emission of CO_2 to the atmosphere.

Appendix J

© Springer Nature Switzerland AG 2021
M. Aresta and A. Dibenedetto, *The Carbon Dioxide Revolution*,
https://doi.org/10.1007/978-3-030-59061-1

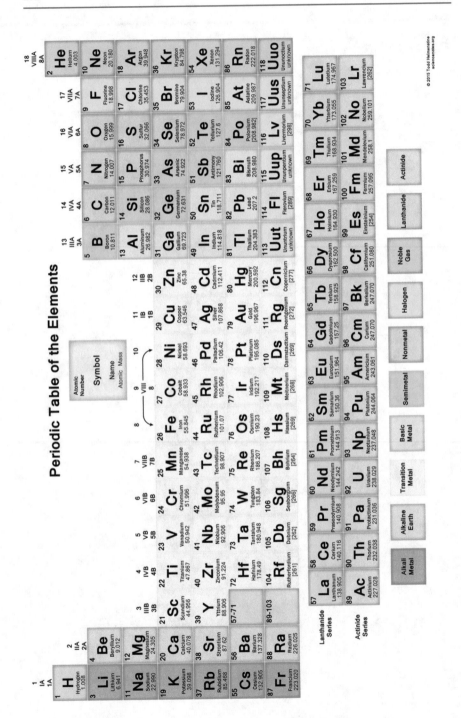

Periodic Table of the Elements

Printed in the United States
by Baker & Taylor Publisher Services